Bill Nye

**ALSO BY BILL NYE**

*Undeniable: Evolution and the Science of Creation*

# UNSTOPPABLE

# UNSTOPPABLE

Harnessing Science to Change the World

*by*

BILL NYE

*Edited by*

COREY S. POWELL

ST. MARTIN'S PRESS ❧ NEW YORK

*To the Next Great Generation.*
*Embrace science. Solve problems. Make things.*
*Change the world.*

www.stmartins.com

Designed by Omar Chapa

The Library of Congress Cataloging-in-Publication Data is available upon request.

ISBN 978-1-250-00714-8 (hardcover)
ISBN 978-1-4668-6989-9 (e-book)

Our books may be purchased in bulk for promotional, educational, or business use. Please contact your local bookseller or the Macmillan Corporate and Premium Sales Department at (800) 221-7945, extension 5442, or by e-mail at MacmillanSpecialMarkets@macmillan.com.

First Edition: November 2015

10  9  8  7  6  5  4  3  2  1

# BILL NYE
## Science Guy®

Bill Nye the Science Guy Show: "Rules"

OBJECTIVE: Change the world.

Produce a TV show that gets kids and adults excited about science, so that the United States will again be the world leader in technology, innovation, and sound mangagement of the environment.

For example, when our audience is of age, we'd like them to produce the best transportation systems in the world, e.g. cars, electric cars, trains, and aircraft.

### Rules of the Road

- The show is entertainment first; curriculum content and presentation of specific facts come later. Ideally, school curricula will follow us.

- All the science we see has to be real science.
  No fictional "molecular resynthesizer" machines that perform magic tricks, for example.

- The science being explored provides the drama. For example, there is no time spent looking for someone's stolen lab coat.

- Science Guy is always himself.
  He could play another character as the Science Guy playing another character. He wears a lab coat and safety glasses for a reason. If he takes them off, it's for a reason.

- Science Guy's reality is television. He can jump from place to place the way a viewer would expect anyone on television to be able to do. There is no need for something like the "Way-Back" machine or the "Transporter" or the "Door to Anywhere." However, the "monitor in the field" can show us supplementary video, e.g. condensation after the walk-in freezer sequence in the pilot.

- Host interacts with guests, kids, other scientists, and celebrities, as peers: E.g. "Hi, Joey; Hi, Michael; Hi, Cindy; Hi Hammer." / "Hi, Bill."

- Show takes place as much as possible in the field. The world is the laboratory.

920601
Revised: 920620/920622/920629/930504

# CONTENTS

# 1

# WE'VE GOT THE WHOLE WORLD IN OUR HANDS

If you like to worry about things—and most people do—you are living at a great time. Climate change is coming, and it is coming right at you. Regardless of where you are on Earth, you will live to see your life or the lives of your kids and their friends change due to the overall warming of the planet. Whether or not those changes are manageable is up to us. It is up to anyone who is able to think about what kind of future we want. It is up to you and me.

I'm sure you've heard people say, "Earth is our home." You may even use the expression yourself. But here is another thought, equally undeniable and even more important: Earth is not just our home, it is also our house. It's our residence, and we are the owners. We are not renters passing through. We are not tenants who can complain to the landlord and eventually move on to live somewhere else. We live here—on this 7,900-mile-wide (13,000-km) ball of rock, water, and air—and we are responsible for its upkeep. Right now, we are doing a pretty bad job as caretakers. We don't seem to be paying anywhere near enough attention to the deteriorating conditions of our home.

Now that you're worrying even a little more than a moment ago, I hope, I'm going to ask you to stop, or at least to move past it. Worry is not going to save us. Neither, by the way, is shooting the messenger (someone like me). I'm asking you to get informed and help fight change with change: change in the way we produce, move, store, and use energy. We can become a great generation that leaves our world— our home—in better shape than it is now while raising the quality of life for people everywhere. This will not be easy. We've already loaded the atmosphere with enough heat-trapping gases of various kinds to cause our planet to keep warming for many, many years to come. But the situation is far from hopeless. Read on, and I will show you why we need to act immediately, what we need to do, and how we can get it done.

There was a moment, a few years ago, when I was really struck by both the true nature of climate change and by the strategy needed to deal with it. I was in Beijing for a meeting of the International Astronautical Congress, a group of rocket people. I observed firsthand a huge environmental upheaval, one of the biggest in this planet's history. Although I was looking right at it, I might have looked right past it, without even recognizing what was happening.

Haosheng Cui, who was a young physics student and a member of The Planetary Society, played tour guide and showed me around Beijing. We had lunch at the famous Qianmen Quanjude Peking Duck restaurant, where an electronic sign announces that they've prepared almost 1¼ billion servings of Peking Duck (they still call it Peking Duck in Beijing). We rode bicycles for the 13-kilometer (8-mile) trip from the conference hotel downtown to the restaurant. Bicycles are still a common way to travel in China, but they are becoming ever less so. Haosheng had an extra bicycle available. It belonged to his father, and his father hardly ever rode it anymore. Their family has become successful enough to own a car.

I couldn't stop thinking about his father's decision to abandon his bike. In a small way, it encapsulated a huge aspect of human nature. We are always looking for ways to do more without having to work so hard. Why bike when you can drive? Why weave by hand, when a machine can do it? Why fight heavy weather and sail with the wind, when an engine can propel your ship? Why ride horses, when a coal- or oil-burning locomotive can take you over a mile a minute? Why travel by train, for that matter, when you can fly in a jet?

That desire—to get more done with less effort—multiplied by billions of people who burn fossil fuels to satisfy that desire, is the root cause of climate change. There are an ever-increasing number of humans on Earth, and every single one of us wants to live a developed-world lifestyle. We want cars instead of bikes. We want electricity that is available any time, day or night. And in the developed world, we're always wanting more: more electronics, more convenience, more luxury. It's an evolutionary impulse to want comfort, to secure as many resources as you can for yourself and your relatives. But the impulse is currently getting us into serious trouble.

Although it all starts with the familiar flames of oil, coal, and natural gas, the details of global warming are complex. I'd say it's like rocket science, but the details of climate change are actually a great deal more complicated than rocket science, by quite a margin. After all, much of our own planet is still a mystery. More than five hundred people have flown in space and twelve people have walked on the moon, but only three humans in history have been to the bottom of the ocean. An orbit in space is clean and predictable, whereas key environmental processes, like the Gulf Stream's interaction with Greenland's ice sheets, are wildly complex. With that said, climate change and rocket science have major things in common: The basics are straightforward, and they're both science. If you have a rocket, you know what to do: Light one end, and point the other end where you want it to go. (Come to think of it—it might be better to point that front end first, and then light the engine on the other end.) In climate science, we can see that we've already lit one end, and we know only too well where it's pointed.

I admit that climate change on a global scale seems hard to believe at first. It's strange that one species out of the sixteen-million-plus on Earth (and a latecomer in evolutionary terms at that) could alter the climate of a whole planet. But it's happening, and we're in the middle of it. We know of only one other species, or group of species, that has the power to change the climate of an entire planet. That would be cyanobacteria, better known as blue-green algae. They were the first organisms to develop photosynthesis, which filled the atmosphere with oxygen and changed the chemistry of everything you see, eat, and breathe. Today that seems great, but billions of years ago, oxygen outright killed things. It wiped out much of the life that could not tolerate oxygen. So yes, there is precedent; one species can change a world. Now we're the ones causing the change, a change that could harm and kill a great many of us. So the question is: What are we going to do about it?

This is when I thought about our planet in a new way. Trading in a car for a bicycle: That is the thinking of a renter who has just moved into a new house or apartment. Developing cleaner, more efficient forms of personal and mass transportation: That is thinking like a homeowner. That is how you take care of things to ensure a long and happy future. Haosheng's dad preferred a car to a bike, especially in bad weather. Why shouldn't he? I have a car, and I find it very useful; he should have that option, too. But so long as we each focus only on our individual decisions and their short-term consequences, we will act like renters, not owners of this Earth. Dealing with climate change requires a new kind of thinking for all of us.

Countless poems, songs, plays, and movies have been written in which the arc of the story shows how love transforms a house into a home. After a few days of backpacking, you might think of your tent as your home, but you would not generally think of your tent the way you do a bricks-and-mortar or lumber-and-siding house. When it comes to your permanent house, you think constantly about its assets and its liabilities, the better to protect and maintain it. The roof is leaking; you repair it. The hot water is not hot; you call a plumber. Should you repaint the outside? Can you do the job yourself or should you hire someone? Can you afford to replace your old windows with better, double-pane windows? Does your house look good from the street? Is that important? Should you spend money on insulation?—and on and on. Your house is your home; it is almost an extension of you, and its well-being is integral to your well-being.

Earth similarly needs our constant attention. We have an atmosphere, an ocean, jungles, deserts, farmland, and cities. Each of these places needs our watchfulness. We have to be careful about what we dump into the atmosphere, because we all breathe and share the same air. We all rely on the same atmosphere for protection. The argument applies equally to our water and land. Some ecosystems are best left alone, but at this stage most of them need our considered maintenance.

How we treat the planet is up to us. We can grow our crops thoughtlessly, or do it in ways that cause little environmental harm. Can we farm with less energy? Can we farm in ways that even contribute to the well-being of the planet? What about our cities? They are often centers of pollution and waste, but they can be centers of innovation as well. They can be the front line for implementing ways to move energy and people more efficiently, and to live well with a lighter planetary footprint.

For most of us, a house is by far the biggest investment we'll make in our lives, often bigger in dollar terms than even the cost of raising a child. It's natural for us to take very good care of our house and treat it as a home. So it should be with our planet. In recent years, you've probably heard a great many people speak about addressing climate change with lists of things we shouldn't be doing—like burning fossil fuels, coal especially. That guidance is useful, but we need to focus more on the things we aggressively *should* be doing—like developing ways to store renewable energy.

Consider a leaking roof. If you're a renter, you put a bucket under the dripping water and you notify the landlord. You might complain about the sluggishness of your landlord's response or the incompetence of the maintenance person, but all in all it's not entirely your problem. You carry on about your business, treating the leak as just an annoyance. If you own the house, a leak quickly becomes a different matter. Many people start in denial. When the rain stops, the leak stops, so you set aside the problem in your mind. You might say to yourself, "It's not that bad; I'll get to it later." But when it happens again—especially if the water is dripping on your television or computer—the leak suddenly becomes the most important thing in your mind. You rearrange the furniture while you're on the phone to the roofer. You get it fixed immediately!

Let's apply that kind of attitude toward climate change. At the moment, a significant number of Earth homeowners are making matters worse, because they're in denial. There are things that we need to take care of right now, a few things that can wait a little while, but all of it must eventually be addressed. Our home needs a good caretaker. Meanwhile, there are more and more of us living in it.

The accelerating increase in the number of humans breathing and burning fuel here on Earth is a huge factor in climate change. When I was eight years old, my family visited New York City to see the 1964 World's Fair. It was a fantastic place. There were dioramas depicting a future that featured cars with rounded aerodynamic shapes quietly slipping through the air on five-level curving superhighways, and forests being felled by clean-running, laser-beaming bulldozers cutting rights-of-way for new highways. There was the enormous stainless steel globe, which still stands in Flushing Meadows near LaGuardia Airport. But looking back, the one exhibit that really got to me was the Population Clock. It was a tote board display that showed the world's human population, and I remember being amazed by how quickly it was growing. A steady drumbeat of advancing numbers indicated how many of us there were; a graph predicted how many of us there would be in coming decades.

I was there with my dad, the same man who felt it was perfectly normal to pull the family car over and creep slowly along the side of the road so he could take pictures (on film, with a macro lens fitted on his Pentax camera) of the car's odometer as it rolled over from 99,999 miles to 100,000 miles. He took a half-dozen shots as the mechanism gently and inexorably nudged the new number and clicked it into place. You can see why a guy like that and his son (shown on the cover of this book) were fascinated by the population-clock tote board. It had the same kind of unstoppable quality.

That day at the World's Fair we were both disappointed when we realized that we'd just missed a monumental flip of the numbers. We arrived too late—by only a few hours—to see the world's population of humans officially rising from 2,999,999,999 to three billion people. Furthermore, I remember signage that described the impending steepening of the global population's growth curve. It gave me pause. It still does. Just since the 1964 World's Fair, the world's population has substantially more than doubled—30 percent more than doubled. We have added another 4 billion to the total, and then some.

Think about the size of the environment we 7.3 billion people share. One of the most recognizable images from the space age is Earth as seen from afar, a blue marble suspended in icy blackness. If you are near a computer or tablet or smartphone (and who isn't these days?), pull up an image of our planet from outer space. Look for the atmosphere, and you'll notice you can't see it, not really. It's as though Earth doesn't even have a layer of gas surrounding it. Relatively speaking, the atmosphere is about as thick as a layer of varnish on a standard classroom globe.

I like to put it this way: If we had some sort of extraordinary ladder-car that allowed us to drive straight up at highway speed, we'd be in outer space in less than an hour. We'd be above the breathable part of the atmosphere in just five minutes! The blackness of outer space is barely a hundred kilometers, or 62 miles, from here, where you and I live. That's it. Earth's atmosphere is very, very thin. And there are 7.3 billion people living in it, breathing it, depending on it, and dumping waste into it.

With every bicycle commuter switching to a car (and with every other energy-consuming "improvement" in our standard of living) we are doing more living, and using more electricity from fossil fuel–powered power plants, and dumping more exhaust gases into the sky. And with every increase in the population there are even more of us doing these things. Population increase is not going to stop anytime soon, and the desire for a better lifestyle is probably not going away, ever. That's why there is climate change. That's why we are facing daunting new patterns of drought, flooding, heat waves, and rising sea levels.

Too often I hear people in the throes of giving up. Climate change is such a big problem, they sigh, that there's nothing we can do to offset it, so we should just let it happen. Let the planet change and we'll figure out how to deal with the consequences. That is the attitude of someone who regards Earth as a rental house, not a permanent home. The problem is, there is no next place to go when the lease is up.

My travels to places like China, India, and Iowa get me fired up about fighting that kind of resignation. Climate change started with

the rise of modern industry. We see its imprint in almost every manufactured product. With each retired bicycle, with each new huge
house built on a deliberately isolated cul-de-sac, with each oversized
air conditioner, and with each jet plane ride for business or pleasure,
we see choices that lead to inefficiency. Fighting climate change can
happen in the same way, by making different choices, by making everything we do—big and small—cleaner and smarter and more efficient.
It's a daunting challenge, but an exciting one. We, as a global generation,
can tackle it.

Our planet may seem huge, but ours is a small world, really,
especially when you look at it power plant by power plant, or even
bicycle by bicycle. Earth is small, a cozy little home, and its future is
in our hands.

# 2

# THE CALL TO GREATNESS

"The climate has always changed in the past and it will always change in the future. We just need to live with it." I hear that sentiment, and variations on it, all the time. I also frequently hear people say, "We have to save Earth." The two things sound like opposites, but they are actually two versions of the same colossal misconception that our primary concern is the well-being of the planet. We don't need to save Earth; it is going to be here no matter what we do. And, we can't ignore the way Earth is changing, because we stubbornly happen to live here.

I encourage everyone to reject both of those sentiments and think instead, "We have to save Earth—for us! For us humans!"

Thinking of Earth as home is a shorthand way of avoiding both kinds of follies. If your house is on fire, you don't comfort yourself with the thought that houses have been catching fire for thousands of years. You don't sit back idly and think, "Oh well, that is the way of nature." You get going, immediately. And you don't spring into action because of an idealistic notion that houses deserve to be saved. You do it because if you don't, you won't have a place to live. If your house is on fire, you are going to work as fast as you can to put that fire out and save your possessions and your loved ones. You probably aren't that concerned about any visiting rats, or bedbugs, or even prized houseplants. You are saving your house for you.

I don't see any single idea or technology that's going to save us from ourselves. Instead, I strongly believe that we are going to have to work at a whole host of problems, in many different ways, all at the same time. We have to create new sources of energy, new ways to store that energy, new ways to transmit it, and new ways to include governments and citizens everywhere. In parallel, we have to overcome the inanity of climate denial. The deniers have wrought havoc. They're standing outside looking right at the burning house and insisting that the yellow flames and dense smoke are not really coming from a fire. It's just the way the house looks this time of day. . . .

For any climate-change doubters who might be reading along, or immediately setting this book down (in a non-recyclable waste bin), let me ask you this: Would you trust a scientist or a politician who insisted, pounding his fist on the table, that there is no connection between smoking cigarettes and lung cancer? These days, probably not. Lung cancer rates went up in lockstep with cigarette use, and the scientific mechanism linking the two was well established. Yet for decades after the science was settled, a great many people, egged on by politicians and industry leaders, doubted this connection.

So it goes again in the current arguments, or "debate," concerning the reality of climate change. The warming of our world has gone in lockstep with the increasing amounts of carbon dioxide and other heat-trapping gases we have dumped into the atmosphere. The connection between climate change and human activity is akin to the connection between cancer and smoking. To wit, it's absolutely certain—for sure. Unlike smoking, though, there is no way for any one of us to opt out. There is no option to quit, or to avoid the risk factor entirely, or to move away from a hard-puffing neighbor. There is nowhere else to go. None of us can leave the neighborhood. When it comes to climate change, we are all in the same house. Earth is where we make our stand. We are all in this together.

You can tell just how serious climate change is by looking at the actions of groups that do not have the luxury of ignoring the gravity of the situation. The United States Department of Defense makes its position clear in a policy statement published in 2014. The document begins:

> Among the future trends that will impact our national security is climate change. Rising global temperatures, changing precipitation patterns, climbing sea levels, and more extreme weather events will intensify the challenges of global instability, hunger, poverty, and conflict. They will likely lead to food and water shortages, pandemic disease, disputes over refugees and resources, and destruction by natural disasters in regions across the globe. In our defense strategy, we refer to climate change as a "threat multiplier" because it has the potential to exacerbate many of the challenges we are dealing with today—from infectious disease to terrorism. We are already beginning to see some of these impacts.

These are not the words of a left-wing pundit. This is the U.S. military talking. They are getting ready for serious troubles ahead. Our house is going to have to be put up on stilts, and we're going to have to secure new supplies of food and drinkable water, and the means to get them to everyone inside. The military is not alone. Insurance companies are taking climate change seriously. Agriculture businesses are taking it seriously. We all will take it seriously, sooner or later. Here's hoping for sooner.

It is an irrefutable feature of our world that everything each and every one of us does affects everyone else, everywhere, because we all share the same air. Little things like recycling your junk mail, installing energy-efficient lightbulbs, and reusing grocery bags, all make a difference, albeit a seemingly small one. But along with the small changes, making a better future is going to require huge ideas and huge actions. We need to think big, because we are going to have to take big steps as a society. As Rick Smalley, the Nobel Prize–winning chemist, put it: "We have to do more with less." We have to provide more food, more water, and more energy to more people, using not just less of Earth's resources—not just less fossil fuel, but no fossil fuel at all. We need to break free of our carbon shackles. I'm sure that if we understand energy and how its production affects the atmosphere, we can do all that. We can engineer a better future.

Before I was known as The Science Guy I spent many years working as an engineer at a few different aerospace firms and one shipyard in the Seattle area. I still have a professional engineering license. Engineers use science every day to build things and to solve problems, sometimes seemingly intractable problems. I'm sure that at some point you've had an idea for something you could build and then managed to bring that idea to fruition. You might have imagined a Halloween costume, created it, and then gotten the reaction you wanted from people. You've probably even surprised yourself a few times by accomplishing more than you thought possible. Having an idea for something and then seeing it built and work the way you planned brings great joy. Now look around you, where you're sitting or standing or lying at this moment. Virtually everything that you see came out of someone's head. It is all the product of science, engineering, and design. Even if you're sitting outside reading, you're probably in a park, and each of those trees was planted and nurtured by people directed by a purposeful plan.

Our society has had enormous success using science to solve problems and make things. Compare how comfortable a modern house can be to what served for housing just a few years ago during the Great Depression. Today, almost everyone in the developed world has clean water, a healthy sanitation system, plenty of food, a climate-controlled home, and seemingly limitless electricity. Compared to any other time in human history, a large fraction of us have an exceptional quality of life. When I look at all that we humans have managed to create, the challenge of climate change does not seem overwhelming. Challenging, to be sure, but not intractable.

Making a better world will involve a great many engineering problems, and every one of these problems will lead us to create a variety of solutions. Even today, there are effective ways to make electricity that create far less pollution than by burning coal. There will be ways to transmit electricity far more efficiently than by pushing it through copper or aluminum wires. There are experimental battery technologies emerging from several labs that could make solar and wind energy vastly more practical, both night and day. There are ways to draw abundant drinking water from the ocean, relieving our direct reliance on snowpacks and rainfall. You and I can help make those things happen.

As an engineer and a tinkerer, I pretty much always look for a technical solution to any problem. I don't just talk about these solutions; I try them out for myself and see how (or if) they work. I treat my own house as a laboratory for living in a better, more efficient way. I'll tell you a lot more about that later in the book. But let me be clear: I do not think we can make our way through the coming decades with technology alone. We are going to have to change the way we manage energy production, transmission, and storage. Those changes cannot just come out of a lab. We are going to need new laws and regulations. I'm not talking about slowing down the economy. On the contrary, I am talking about doing more with the resources we have, by paying attention to where our energy comes from and where it gets used.

The challenge of climate change reminds me of when I was a kid vacationing with my family along the Delaware shore. Every now and then, people would stop in their sandy tracks and stare out past the breakers. They had noticed an Atlantic bottlenose dolphin swimming by. On extraordinary days, they might see half a dozen animals gliding along and arcing above the surface for a graceful moment to breathe. Soon people up and down the beach in the sleepy resort town would point and shout: "A dolphin! Look! Wow, that is fantastic." But this situation has changed, and in a good way. Nowadays, it's not unusual to see hundreds of dolphins every day. We changed the dolphins' world; we cleaned it up.

Back when I was young, it was legal for ship crews to wash out their bilges with seawater. Wastewater and rainfall collects inside the ship at the bottom of the hull; that's also where any leaked oil from the engines or galley drains. Ships used to rinse their insides with seawater and dump the resulting bilge directly into the ocean, into the dolphin's habitat. The problem was so bad that when I returned from the beach each afternoon, my feet would be smeared with black goo: tar from the bilgewater that washed up on shore. Families would keep gasoline-filled lids outside the front door of their cottages, along with an old rag. Before being allowed in the house, I had to dip the rag in the gasoline and wipe the tar off my feet. There was tar on the sand on pretty much every beach up and down the Delaware and New Jersey shores. Naturally, it ended up on the floors of houses, too.

That's just the way it was, but it's not the way things are anymore. We stopped accepting long streaks of tar on our beaches. Lawmakers, elected by people like you and me, put an end to it. Crews were no longer allowed to dump thick old oil and trash out of the bilges of passing cargo ships. We made a mess, but then we cleaned it up. We changed the world.

My parents' generation started out working hard, focused on earning a living without much concern about creating trouble in the environment. They weren't out to harm the planet; they simply were not aware of what they were doing, or of how they might do it differently. In the case of the Eastern Shore beaches, the old approach left a thick mess on the places they took their families to play. After some time, they came to see that the long-term quality of the environment is a more worthy priority than the short-term need to rinse out the insides of ship hulls cheap and easy. They realized that change was possible, that change was not even all that hard or expensive, and that one small improvement could make a big difference. They saw to it that the local waters got cleaned up. They left the beach a little better than they found it.

There have been a great many books, and articles, and documentaries about the World War II generation. People like my parents, both veterans of World War II, came to be called the "Greatest Generation," because they rose to the challenge and defended the world against tyranny. Often enough, certain pundits imply that no generation since—today's generation, especially—can live up to the standard of the greatest generation. I could not disagree more. We face a challenge right now, you and I, that is even greater in aspect and scope than a global war. It is a battle for our house and home, and for our future on this planet. It is a moment for all of us to step up: through our personal effort, through the innovations we create, through the policies we support, through the people we vote for.

You and I can be a part of the Next Great Generation. We can save Earth—for us. Let's get to work.

# 3

# A HOTHOUSE OF DISBELIEF

When I was in engineering school at Cornell University, I had the wonderful good fortune of studying astronomy with Carl Sagan. He was among those few accomplished scientists who insisted on teaching the entry-level courses, the 101s and the 102s. He didn't see a wall between the ideas he explored in his research and the ideas he wanted to share with every student—even the ones who had, until then, had little serious exposure to science.

As citizens of a warming world, we have a great many tasks before us. While we're establishing a renewable electrical grid and providing clean water for all of the world's people, dealing with climate-change deniers is just one more job that has to get done. I often wonder how Dr. Sagan would be reacting to the strident deniers if he were still alive. He died in 1996, before the many consecutive floods in the U.S. Midwest and the extraordinary drought conditions in California kicked in. I am sure he would have been vocal and influential, encouraging us to address climate change as quickly as possible. Decades before most of us, he recognized the implications of a warmer world. Here's hoping climate-change deniers today take a little time to hear what Sagan had to say then.

One of the most innovative areas of Sagan's research concerned the way in which planetary atmospheres trap solar energy as heat. Along with a group of collaborators, especially NASA astrophysicist James Pollack, Sagan developed a computer model that described the warming of our world caused by this process—the oft-cited greenhouse effect. Energy from the Sun, mostly in the form of visible light, passes through the gases of the atmosphere, strikes the planet's surface, and is partially absorbed. The surface warms a bit. The warmed surface partially reradiates some of this energy, as heat. On its way back out to space some of the heat energy is trapped by certain atmospheric gases, especially water vapor and carbon oxide.

A carbon dioxide molecule is linear. It's an atom of oxygen connected to an atom of carbon, connected to an atom of oxygen, all in a row. It's the right length and of the right atomic flexibility (or floppiness) to allow visible light, with wavelengths ranging between about 390 and 700 nanometers (billionths of a meter), to pass right by. But, these molecules block the longer reradiated infrared rays (heat), whose wavelengths are about ten times as long as those of visible light. That heat-trapping ability is a feature of the size and shape of carbon dioxide molecules, and the length of the waves they trap or let pass. Yes, this really is somewhat like what happens in a greenhouse. A greenhouse traps warm air inside that's generated by the transpiration of photosynthesizing plants. The warm air does not get driven up and away by convection or wind. The glass of the greenhouse, like carbon dioxide, is transparent to light but does a pretty good job of blocking heat. Solar energy comes in, but a lot of it does not get back out. That is how a greenhouse can stay warm even in winter.

Without our atmosphere holding in heat through the greenhouse effect, Earth would be completely inhospitable for us. It would be a frozen world, with an average temperature of about −18°C—or just under 0°F. The problem is not that we have a greenhouse effect; the problem is that our greenhouse effect is getting stronger by the hour. The extra heat in the atmosphere is altering weather patterns and local climates around the world. To be sure, there are many things that influence the climate beside the greenhouse effect. When volcanoes erupt, they generally cool the world a little because the dust and aerosols they blast to high altitudes reflect sunlight back into space. The sun itself varies in brightness a tiny bit from year to year. But the kinds of atmospheric models that evolved from those developed by Pollack and Sagan, combined with a tremendous amount of environmental and historical data, show that carbon dioxide—produced by us burning fossil fuels—is causing the overwhelming majority of the current climate change.

And yet, even now, after nearly fifty years of intensive research and more than thirty years of scientific consensus on the nature of global warming, there is still a sizeable group of people who grasp at any stray fact or, more commonly, any stray intuition, to help them deny what is happening. In denying the problem, they are also denying the need to step up and do something about it.

Look, I get it. The magnitude of climate change can be really hard to grasp, because it often runs counter to our intuition. If you're from the state of Oklahoma, and you grew up in a rural area, where your nearest neighbor was miles away, you probably cannot imagine how a few people spread out over a wide prairie could possibly influence the climate of an entire planet. Nevertheless, climate change is happening, and in different ways in different places. Decades-old personal experience is not a useful guide for evaluating what is happening on this grand scale.

I know firsthand how misleading intuition can be. When I was in high school, my friends and I celebrated a few especially cold winters by ice-skating and playing hockey by lantern light on the Chesapeake and Ohio (C&O) Canal, which runs right along the Potomac River in Washington, D.C. Those memories really stood out. So it seemed reasonable to me in 1975 when *Newsweek* reported that some scientists believed the world was cooling, and we were headed for another ice age in the coming centuries. Then as now, *Newsweek* was not a scientific publication. It was in the business of selling magazines—in part perhaps by making stories sound a little bigger and scarier than they really are. At the time, I didn't know that the U.S. National Academy of Sciences also published a report in 1975, concluding that "we do not have a good quantitative understanding of our climate machine and what determines its course. Without the fundamental understanding, it does not seem possible to predict climate."

But I was a teenager, and that was forty years ago—when computers were primitive, when the Internet connected just a handful of research centers, when Earth-observing satellites were few and limited. Today we know vastly more about how the climate works. Nevertheless, climate deniers routinely cite that 1975 *Newsweek* article and a similar one from *The New York Times* as authoritative sources proving that scientists have no idea how to understand climate. Anyone who insists on citing these old publications today is not just misunderstanding the science; he or she is ignoring reality, and trying to get you to do the same. There is a political agenda at work, fueled with money. Don't fall for it, please.

There are several other arguments that the climate-change deniers use continually. A common thread is that they sound reasonable, until you really examine them closely. Join me for a few moments while I do just that. Picking apart the flaws in these arguments is a great way to learn more about the real science of climate change, and that in turn will give you the kind of knowledge that might help open up a few minds. As I say, denialism is a problem that needs to be solved in parallel with providing clean energy and water to the world's humans.

People who should (or do) know better keep confusing weather with climate. Weather is what happens day to day in one place. Climate is what happens over many years to a large geographic area, or the planet as a whole. In February of 2015, Senator James Inhofe brought a snowball to the Senate floor to demonstrate that global warming must be a hoax, because it snowed in Washington, D.C.—in winter. Along with inane conflating of weather and climate, the senator also showed profound misunderstanding of how climate change operates. Not every place will become warmer all the time; climate change is much more complicated than that (which is why many scientists have cut back on using the shorthand expression "global warming"). With the western portion of the North Atlantic slightly warmer than usual, you would expect more moisture in the atmosphere and more snow along the eastern seaboard of North America. There is currently no way to say that global warming caused one specific snowfall, but more energy in the atmosphere increased the odds of a storm like the one that struck Washington, D.C. The storm certainly did not discredit scientists' climate models, as the senator claimed.

From time to time, an ill-informed person or someone intent on deliberately misleading his or her audience will opine that having more carbon dioxide in the atmosphere could be a good thing, because plants metabolize carbon dioxide. Carbon dioxide provides carbon to plants, true enough. It also makes our world habitably warm. The problem isn't that there's carbon dioxide in the atmosphere; the problem is the rate or the speed at which we are adding more greenhouse gases, carbon dioxide especially. Since 1750, humans have increased the amount of carbon dioxide in the atmosphere by 40 percent, to more than 400 parts per million.

Adapting to the rapid changes we're inducing will take a great deal of time. Meanwhile, many areas will become too warm or arid for the plants that currently live there. In the near term, persistent droughts over the same areas year after year will expand our existing deserts. California is just beginning to feel the effects of such extreme, prolonged drying. That, along with the expanding ocean and the concomitant rise of sea levels, will cause the amount of green land to shrink, not expand, as the atmosphere's carbon dioxide level rises. There are plenty of green plants in the sea, but the scale of the changes will cause trouble for them as well. Both land and sea plants are adapted to their local climates, the conditions that we have now. Rapid climate changes will harm or kill a lot of them. All of this is bad news for humans. Climate change and the resulting rise in sea level will also leave our coastal cities vulnerable to flooding. So, the more-$CO_2$-is-good argument is just silly.

Sometimes climate-change deniers will point to the Medieval Warm Period, a time when some parts of the world, notably the North Atlantic, got a little bit warmer from about the year 950 to the year 1250. This was about the same time that Scandinavian settlers made their way to what came to be called Greenland. Climate deniers take this as proof that Earth's climate varies quite a bit on its own. They conclude that since this happened once before, humans are not affecting the climate today, that natural variation is so great that there's no way humans can be held responsible. Of course, scientists work hard to take every natural input into account, when they mathematically model how the climate is changing. Diligent climate scientists have carefully assessed the natural processes involved and shown that what is happening now is unprecedented. It's completely different from the Medieval Warm Period. Using locally documented warm spells from the distant past as proof that global human-caused climate change does not exist today is the kind of superficially convincing claim that is useful for confusing television viewers or scoring political points, but it is not scientifically meaningful.

Another spurious and distracting use of a single datum is the observation that the atmosphere has not warmed as much over the past decade as computer models predicted. Deniers conclude that the models must be completely wrong because of this apparent discrepancy. A recent analysis by scientists at NOAA (the National Oceanic and Atmospheric Administration) shows that, when you look at all the data, the world is actually *warmer* than the models said it would be. The discrepancies are in the details. The amount of heat stored in the atmosphere was not quite as high as was predicted. The data were reviewed. The temperature readings were sorted. It turned out that the theretofore-missing heat ended up in the ocean, producing more deep-sea warming than originally expected. Furthermore, since the models were run, the Sun has been slightly less active than normal, and an episodic shift in ocean currents (called La Niña) made the surface of the Pacific Ocean a little cooler than it is on average. Such natural variations have virtually no effect on the long-term trend, but they can produce deceptive short-term patterns. Meanwhile, the warmer-than-expected sea has accelerated the melting of Arctic sea ice. The much-hyped "pause" in global warming exists only if you cherry-pick certain details and ignore all the rest.

Perhaps the most intense line of denialist attack goes after the climate models themselves: Scientists can't even predict next week's weather, so how can they be sure what next century's climate will be like? It's the bigger version of the argument about the *Newsweek* article. Basically the question boils down to: Why should we believe that these climate scientists know what they are talking about?

It's a good question, actually, and one that's worth answering in some detail. Modern climate models are extremely complicated, much more so than the ones Pollack and Sagan were using back in my student days. Today's models include the effects of clouds, oceans, glaciers, mountains, forests, open lands, and all of the different gases and particles that humans are sending into the atmosphere. As Earth circles (or ellipses, *sic*) the Sun, its axis wobbles a little on a twenty-six-thousand-year cycle. Earth's orbit itself oscillates, changing its ellipticity over a one-hundred-thousand-year cycle, and it shifts its tilt over a forty-one-thousand-year cycle. These motions change the amount of sunlight that falls on different parts of Earth at the different times of year and over millennia. To be accurate, the models need to include all of these variations—and they do.

The results are computer models that represent Earth, its atmosphere, and its seas, using everything we know. Mathematicians and scientists use the physics and chemistry of water, air, and sunlight to create billions of imaginary boxes in the sky. They mathematically simulate air, water, and land. The individual properties of each box—each mathematical element—are well known. The researchers build complex equations to describe how each box interacts with all the boxes around it to determine what will probably happen next, i.e., as time goes by. But just as important, they check the models by using them to predict what has already happened; they run the models backward and "postdict" the past. It's a term of art closely related to terms of sport, especially the mathematics of baseball.

One of the fascinating things about baseball compared with a great many other sports is that in baseball, the statistics are especially meaningful. Each pitch is a datum. Each outcome of that pitch, whether it be a ball, a strike, a hit that takes the runner one, two, or three bases, or all the way around to home plate, goes in the baseball record books. In recent years, those statistics have been revisited to come up with a set of criteria that can, as the saying goes, "predict the past." Modern baseball statisticians have revolutionized the way players are chosen and used in each game by comparing what has happened with each player's past at-bats with the overall outcome for each team each season. This technique was celebrated in the book and the movie *Moneyball*. Any proposed set of statistics is not considered to be of any special value unless it can be used to show that outcomes of a past season are consistent with what these statistics would have predicted had they been in use years ago.

So it is with climate models. We want to put certain data into a complex computer program and have the outcome be consistent with climate data gathered from patterns in rocks, pollen grains found in lake sediments, tree ring counts, and the bubbles of ancient atmosphere trapped in the ice found in Greenland, Antarctica, and elsewhere. A climate computer model is not trusted unless it can predict the past. Nowadays, the models are converging. We are getting increasingly accurate and we are seeing exactly how serious climate change will be.

Statistics also explain why it is much easier to predict big, long-term climate trends than it is to predict small, local weather events. On the short term, you are trying to follow what is basically a chaotic pattern. On the long term, you are tracking the overall effect of changing inputs. With the possible exception of senators from Oklahoma, nobody questions you if you predict that summer in the United States will be warmer than winter: Earth's tilt ensures that there is more sunshine in a given hemisphere during its summer, and so surface temperatures there are higher. There can be cool summer days and warm winter days, but overall summer will always be warmer. Asking how scientists can be sure the planet will keep getting warmer is like asking how scientists can be sure that summer will come next year. It's all about the big trends.

Speaking of trends, when I started doing the *Bill Nye the Science Guy* television show in 1992, our atmosphere had 356 parts per million of carbon dioxide. Today, as I mentioned, we are over 400 ppm. That's all from us and our fossil-fuel burning. The more greenhouse gas we create, the more warming we will get today, and the more warming we will continue to get in the coming decades and centuries. To make the situation even more serious and the need for action even more apparent and urgent, bear in mind that there is no stopping a large fraction of future warming, because billions and billions of tons of the gases that are going to bring it on are already in the air. Even the most fragile of them do not break down for decades. Their effects will be felt for millennia to come.

I emphasize that carbon dioxide is very significant, but it is not the only greenhouse gas. Human activity also produces methane. Because of the characteristics of methane molecules, they hold in heat much more effectively than carbon dioxide molecules. And methane molecules remain intact for decades. Over a century timescale, a kilogram of methane is about 30 times as powerful as a kilogram of carbon dioxide. Now get this. There are billions of tons of methane in molecular cages of water ice, called "clathrates," held tight in the permafrost soil of Earth's northern regions and at the cold bottom of the ocean. As Siberia warms, and as the water that circulates along the seafloor warms, the clathrates in the sediments both on land and deep in the sea will release the trapped methane molecules. Once liberated, they'll come bubbling up. The methane gas will come out in the same fashion that bubbles are released by opening a bottle of soda, or when you pop the cork on a bottle of champagne.

Scientists don't know yet how much methane clathrates will add to the warming process. But intuitively, when one considers how much permafrost there is, or used to be, it seems likely that there's a lot of methane and a lot of potential for a lot of trouble. More important, the recent research into the possible impact of clathrates makes a crucial point: Yes, there are uncertainties in the climate projections, but many of those uncertainties are things that would make the warming much worse than restrained scientists are forecasting.

So why are we, any of us, still debating the reality of climate change? Why aren't all of our political and business leaders joining the cry to rally the Next Greatest Generation to come up with some solutions? A key part of the problem is that many of our richest people made their fortunes in the fossil-fuel industry. To protect their wealth and businesses, they have turned to promoting denial. Conservative politicians get a great deal of their campaign contributions from fossil-fuel wealth, and they have been convinced to interchange the standard statements of scientific uncertainty (e.g. "plus-or-minus 3%") to mean that we know nothing at all (i.e. "maybe the answer is minus 100%"). Conservative media outlets have obediently played along. This is wrong and dangerous.

By way of example, I'd like to share my recent experience with one of the producers working with John Stossel, who was once a mainstream commentator on ABC News here in the United States, but later moved to Fox News in 2009. He's generally very critical of progressive politicians, and he's an ardent climate-change denier. One of Stossel's producers sent my publicist an e-mail asking why anyone should give my commentary about climate change any more credence than anyone else's commentary. My publicist and I were further charmed by the inclusion of a deadline for a response. In so many words the producer said: "You have until Monday noon to respond to this note." Hers was an unsolicited request for information that is available online, so having a deadline added a confrontational tone. It struck me as a challenge, and one that I would enjoy taking on.

For me, this was an opportunity very much like my choosing to participate in a creationism debate in Kentucky in 2014. I am inclined to meet the lion—or in this case, the Fox—in his den. I feel that exposing the extraordinary denialism that Fox News promotes, to the very audience that does not question it, may be an effective way to change minds, albeit slowly. I often point out that one debate on the Internet or one television interview is not going to change anyone's perception like a toggle switch: one way, then the other way. Instead, it's a process. My appearing on the Fox Business Network is one more chip knocked out of the wall of thoughtless climate denial. Here's what I wrote back:

*Dear Ms. [Producer] et al,*

*I am a mechanical engineer, Washington State [P.E.] License #21531. I became an informal science educator after more than twenty years in aerospace in the Seattle area. As an engineer, I studied physics, classical physics especially for four years in school. I took courses in heat transfer and fluid mechanics. The physics of the atmosphere involve both. I made a living as a specialist in mechanical design and what's called "control theory," the niche of engineering concerned with feedback systems, e.g. aircraft control surfaces and business jet autopilots. The atmosphere has feedbacks and forcings as well.*

*I chose to work to influence kids, because of what I felt was a pattern of mediocre management practices in the U.S. The trend manifested itself in the production of the Ford Pinto, Chevy Vega, the removal of the solar hot water system from the White House, the neglect of the metric system, which retarded our competitiveness internationally, the destruction of the General Motors EV-1 electric automobile, and the billions lost in the development of warplanes that either could not fly or are renowned not for their performance, but for their exorbitant cost.*

*That aside, I cannot help but answer a question with a question. What on Earth makes anyone think that 97% of the world's climate scientists are wrong? I'm not asking about one or two articles in popular magazines in the 1970s. I'm asking about the extensive peer reviewed publications available today. I can ask the same question in another more personal fashion, is Mr. Stossel in the first stage of grieving about the Earth's environment, i.e. is he in denial? The next four stages of his grief would then include anger, bargaining, depression, and acceptance, per the Kübler-Ross model of grief.*

*Fox viewers should keep in mind that in the science and engineering communities, we have been very concerned about climate change since the 1980s. You may recall 23 June 1988, the day that Jim Hansen testified before the U.S. Congress about climate change. I presented demonstrations about climate change on my kids show* Bill Nye the Science Guy *in the 1990s. I mentioned it in my first kids book* Bill Nye the Science Guy's Big Blast of Science, *in 1993. The discovery of climate change, and our society's reaction to the discovery induced Mr. Stossel's behated (sic) former Vice President Gore to produce his movie* An Inconvenient Truth *almost 10 years ago.*

*We on the outside have moved to acceptance of the phenomenon of climate change, and we are working to do something about it. Mr. Stossel could influence his viewers so that they join the rest of the world and get to work addressing [this issue].*

*Onward,*
*Bill Nye*

After receiving no reply from Mr. Stossel's producer, I added this a few days later:

*Also for Mr. Stossel's and your consideration is the attached Rules of the Road document for the* Bill Nye, the Science Guy *show. I wrote this single page document and gave it to each of the dozens of employees and interns, who worked on the show. It reflects my concern about the environment and the state of U.S. engineering management back then, a concern I still have, of course. Note that it was created on 1 June 1992, with a final release date of 4 May 1993. Other examples of mediocrity for me include the Challenger and Columbia Space Shuttle wrecks. Also as you may know, I spent two years in the oil industry. I worked as engineer on skimming oil slicks and on a machine to remove tar and paraffin-bearing water produced with old oil wells in Texas and New Mexico. I washed my coveralls in laundromat greaser machines; I learned a little bit about oil production.*

John Stossel seemed to have lost his nerve; he never did have me on his show. But I hope he's willing to consider that even a staunch libertarian should be in favor of dealing with climate change. Those who are contributing to the problem are impinging on the right of all of us to enjoy Earth the way it is now. Sometimes political freedom needs defending. The same is true of environmental freedom. If we want balanced ecosystems everywhere on Earth, we are going to have to come up with new ideas and do new things on enormous scales. We can do it, but only if we stop denying the problem.

# 4

# PUTTING A PRICE ON INACTION

Being a homeowner is a constant act of weighing costs and benefits: What investments do you really need for your home to be safe and secure? Do you need smoke detectors, a radon measurement, a dehumidifier, new insulation, asbestos removal? So it is with our planetary home. Every new technology or new behavior that will help address climate change takes effort, and most of them require investment as well. That simple truth has led to a whole second category of climate denial. Many people think that change is coming, but that dealing with it is more trouble than it's worth, too difficult and expensive. I'm here to tell ya', they are dead wrong.

Here's a rundown of just a few reasons why climate change and the warming of our world is such a serious business. Try this: Fill a glass with water. Measure the water level as carefully as you can and put a mark there; use a piece of tape and a pen or pencil, or maybe a felt-tipped marker. Put the glass of water in a microwave and run it, just for a minute or so. Look very carefully at the mark on the tape. The water level will have gone up—just a little. So it is with the world's ocean. As it gets warmer—just a little bit warmer—it will get bigger, just a little. Water expands when warmed. The warming of seawater is a major cause of our rising sea levels. Melting glaciers and ice sheets also contribute to the rise. As the ocean expands, it will overrun seaports around the world. Wharves and streets that provide access to the cargo that comes and goes are just a meter or two, a few feet, above the level of the sea surface.

Now, think about cities you may know: New Orleans, Miami, San Francisco, Seattle, San Diego, Los Angeles, and New York. Think about other coastal cities around the world: Tokyo, Sydney, Mumbai, and Qingdao. As the climate continues to change, those places and the surrounding areas are going to have to build seawalls and dykes to control flooding. At first, they'll be flooded only when the tide is especially high or when storm winds drag water up into a "surge." But as the decades pass, the same coastal areas will get flooded all the time, and people will be forced to abandon their homes and move . . . somewhere. All that infrastructure and all that investment— the pipes, the wires, the lumber, the roof shingles, and the sewers— will be left behind. If you live in the developing world, it's going to be a real mess, because generally you don't have another inland area to move to. You're stuck. . . . And it's largely the developed world's fault. That's where most of this greenhouse gas was produced, yet you in the developing world will have to deal with it in the worst way. Who can blame you for your anger?

With more heat energy in the atmosphere, storms are going to get stronger. If the physics and heat-transport mechanisms of an individual storm cause it to spread out rather than intensify, it's still trouble. With more heat energy in the atmosphere, there will be more water carried aloft as vapor which will fall as precipitation—and that will cause trouble. By that I mean, almost every bit of infrastructure we've created in the United States over, say, the last ten decades is not suited to such large-scale changes in weather patterns. We will be literally out of our depth. The same is true all over the world. The 2015 heat waves in India and Pakistan killed roughly four thousand people. That's just one small hint of what is on the way.

As I write, I routinely post messages on social media pointing out the intensity of rainfall and frequency in flooding here in the central part of the U.S. We didn't use to have tornadoes in Chicago or snow-drifts over the roofs of buildings in Boston for months at a time. There is no way to connect any one event to climate change, but overall these kinds of extreme-weather events will become more common. It's not just inconvenient. Right here in California, where I live part time, we have had years of drought. Governor Brown has had to institute radical cuts in how water is allocated, and they may still not be enough. California's $40 billion agriculture industry is in crisis. There is not enough winter snow to fall in the mountains to supply the cities of the West Coast, as well as the crops there, crops that the U.S. and a great many in other parts of the world depend on.

At the North and South poles we're headed for trouble in two ways. In the Arctic Ocean, the ice floats on the sea and reflects sun-light into space. As it melts, the sea would stay the same level—like what happens in a glass of ice water—except, that since it's soaking up sunlight and getting warmer, it's expanding. And as ice melts in Greenland, fresh meltwater is flowing into the surrounding sea, rais-ing its level and making it less salty. The less salty water is not as dense as it used to be, so does not sink at its former rate. The vertical currents have slowed. The net effect is deflecting the Gulf Stream, and that is changing the weather on two continents.

As ice melts at the South Pole, in contrast, huge sheets of frozen water will slide off the rocky continent of Antarctica and fall into the Southern Ocean. It's like dropping an ice cube into a glass of water so that it overflows onto the countertop. These Antarctic ice sheets are massive. The amount of water they contain is huge. When they slide into the ocean, sea levels around the world will go up. There's no way around it. Earth is our great big spherical house. What happens up, down, and over there affects all of us.

Next there is the matter of food. Our crops come from agricultural areas that have been carefully developed and cultivated over the last ten centuries. As we change the pattern of rainfall and the frequency of severe weather events, we are going to have to change where and when we grow our food. North America provides supplemental food supplies to populations around the world. As the weather changes, our farming system is going to have to change as well. It won't be so easy. Right now, we don't have the means or especially the infrastructure to grow the same amount of food in northern Saskatchewan, say, that we currently produce in southern Nebraska. It will take decades to make agricultural changes on such scales.

Climate change will affect disease as well. As I've often remarked, our biggest enemies from an evolutionary standpoint are not lions, and tigers, and bears. They are germs and parasites. As our world warms a little, the ranges over which certain germs and parasites can live continue to expand. Human populations that used to be safe from tropical diseases no longer will be. Pine forests protected by freezing winter temperatures are being destroyed by beetles that no longer have to deal with the cold. There will be enormous, heretofore unexpected costs dealing with the disease, death, lost wages, and lost productivity as more of us get sick more and more frequently. It seems like a subtle effect, but we should all remember the bubonic plague, the so-called Black Death in the sixteenth century, and we should all be very well aware of the Spanish Flu epidemic that killed at least 50 million people in 1918–19, more than the number killed by combat in World War I. A warmer world is good for parasites and bad for us. Malaria and other tropical diseases are already venturing farther north, as a warming climate welcomes them to places where winter frost previously kept them at bay.

This is where I come back to the idea of the house on fire. You can run all of the cost-benefit calculations you like, but when you see those flames the decision is made: You have to act. You can't afford not to. When it comes to safeguarding our planetary home, it's the same deal. We have no option but to act. Fortunately, there is still just enough time—and our science will show the way.

# 5

# INPUTS AND FEEDBACKS

Shopping for music early in the twenty-first century, I cannot help but notice the resurgence of vinyl phonograph records. A very thin gemstone vibrates as it slides through the groove in a spinning vinyl disk, and electronically those vibrations become sound. I am of a certain age; I grew up with these kinds of records. When I was a little kid, barely able to manipulate a seven-inch vinyl "single" and place it on the turntable, there was one song that fascinated me, a double-diamond (20 million+) record: "Sixteen Tons," by Tennessee Ernie Ford. It features an unmistakable, melancholy clarinet riff that sets the scene for Mr. Ford as he sings of a coal miner's lament and the backbreaking work of his daily haul. That is the world of the carbon economy—the world we are trying to change.

Sixteen tons of coal would fill about a dozen of the small railcars used in mines. It's nearly seventeen cubic meters of black rock. Tennessee Ernie Ford took those numbers and turned them into country music poetry. I'm here to turn them into a concrete measure of what today's system of energy is doing to us, and exactly how we need to change it.

A lump of coal is nearly pure carbon. When you burn it, each carbon atom hooks up with two oxygen atoms from the atmosphere to make carbon dioxide, $CO_2$. An oxygen atom weighs one-third more than a carbon atom, so the greenhouse gases add up quickly. When you burn one kilogram of coal, you get 3⅔ kilos of carbon dioxide. At standard atmospheric pressure and temperature, one ton of $CO_2$ takes up 534 cubic meters, which would fill a cube 8.1 meters (almost 27 feet) on a side. That's about as tall as U.S. football goalposts (30 foot minimum). Tennessee Ernie Ford's 16 tons of coal would become about 59 tons of $CO_2$, which would fill a cube 30 meters (100 feet) on a side—enough to fill a good-sized office building. That's a lot of gas, and we're only getting started.

Humans create about 1,150 tons of $CO_2$ every second. That's over 96 million tons of $CO_2$ around the world every day, which in turn adds up to about 35 billion tons a year. Over the last couple of centuries, we have tossed up enough carbon to convert almost 4 trillion tons of life-giving oxygen into 6 trillion tons of carbon dioxide. It's mind-boggling, or it might be more accurate to say it's mind-numbing. We've been combusting so much air for so long that we're all used to it. In American slang, we might try to wave it off by saying, "Ain't no big deal." And for a few centuries, it wasn't such a big deal, but now it is.

Before the steam engine took off around the middle of the eighteenth century, carbon dioxide made up 280 parts per million of Earth's atmosphere. That may not seem like much, but it is enough to sustain every green plant you've ever met, above and below the surface of the sea. Those plants, in turn, sustain all of Earth's animals, including you and me. The carbon cycle as a whole sustains the planet. It's amazing, and when it stays in balance, it is a beautiful thing. Right now it's not in balance, however. The 35 billion tons a year we add to the mix are changing our planet's entire atmosphere. In 2014, carbon dioxide levels worldwide topped 400 parts per million for the first time in human history.

Let's be clear, carbon dioxide is the most important greenhouse gas that persists in the air. It doesn't go up and down like water does as vapor, rain, and snow. It is also deceptively easy to overlook; unlike soot or other smog-style pollution, carbon dioxide is clear, invisible. Nevertheless, our production of it is changing the world. Carbon dioxide is troublesome in two ways. First it really does allow visible light to pass through it, and it really does trap the infrared light or heat coming back up from Earth's surface. It's a global blanket exerting more influence on the world's temperature today than it has in millennia. The second big problem with carbon dioxide is that it lasts and lasts. It persists in the atmosphere for millennia. That's why there is global warming and concomitant climate change already built into the system. The carbon dioxide we've added so far is going to keep warming the world for centuries to come. The last thing we want to do is add even more. We must stop pumping carbon dioxide in the air, the sooner the absolute better. We want to decarburize (*sic*) our economy. We want to get our electricity from other sources without burning old buried fuel.

Back in the good ole cretaceous period, 100 million years ago, in the days of the ancient dinosaurs, there was three times as much carbon dioxide in the air as there is today. And sure enough, there was a lot less ice and snow. And sure enough, the middle of what is now North America was an inland sea. And sure enough, we absolutely cannot let that happen again. If we do, there will be a lot fewer of us alive.

You may notice that life was flourishing back in the cretaceous. Living things were happily adapted to the way things were. Climate deniers sometimes like to throw this point into their bag of arguments as well: See, lots of $CO_2$ isn't so bad after all! Uh yeah, it's fine . . . if you happen to be a dinosaur. Herein lies an essential idea. We cannot remind ourselves often enough: It's not the total amount of carbon dioxide in the air today that poses a threat to us. It is the rate of change—the speed at which the fraction of $CO_2$ is growing—and our modern society is built around a specific kind of climate and a specific level of carbon dioxide.

Historical datasets are crucial for calibrating how the climate responded to changing $CO_2$ levels in the past. That information helps scientists tune their climate models, and understand just how much we need to cut emissions in the future. You might well wonder, how do we know how much carbon dioxide there was in our air two and a half centuries ago? Well, we can directly measure how much $CO_2$ was around tens of thousands, even hundreds of thousands of years ago, because we have discovered places where the ancient atmosphere is preserved: It is locked away in the ice sheets of Antarctica, Greenland, and Siberia.

Here's what happens. When it snows in those locations, the tines (the little fingerlike extensions, like those on a dinner fork) of the snowflakes fall together and trap tiny pockets of air. As it snows and snows, year after year, the tiny pockets betwixt the tines get smooshed, and the snowflakes get bent and broken by the weight of the snow above. As they mechanically deform, the flakes liberate a tiny bit of heat, softening the surrounding crystals, and the whole snowpack turns to often perfectly clear ice with tiny bubbles embedded. The process is often called regelation (the water in the ice "re-gels"). Those trapped bubbles are the gases of the atmosphere at the time the ice formed. When scientists get hold of those bubbles, they can determine with extraordinary precision just what mixture of gases they're dealing with.

Just as beams of light can be separated into specific colors to reveal a spectrum, the masses of gases can be assayed in wonderful machines that shoot atoms and molecules horizontally across a vacuum chamber. By measuring where the atoms fall, scientists can determine their masses and deduce just what quantity of which atoms are present in a sample. Because atoms are so tiny, it doesn't take much of a sample to get a wealth of information. Using these mass spectrometers, scientists have measured the amount of carbon dioxide and other gases in Earth's atmosphere back over hundreds of thousands of years. This technique is extraordinarily accurate and incontrovertible. We can see exactly when the Industrial Revolution started, and we can assay the $CO_2$ content of the atmosphere with precision over decades and centuries.

A few years ago, I visited the U.S. National Ice Core Laboratory in Glendale, Colorado, to see this research for myself. The main building houses thousands of long solid cylinders of nearly clear regelated snow ice, harvested diligently from all over the world. The building is kept at -36°C (-33°F). I asked the head guy, Todd Hinckley, why he chose this particular temperature to store and preserve these hard-to-get, expensively acquired, slugs of ice. He explained that he didn't choose the temperature as such. The building engineers just turned down the thermostat as cold as it would go. And it is cold! To be in that area, you have to put on super-boots and layer upon layer of garments, and that's just to hang in there for a few minutes. Whew, it's cold in there!

Digging into the history of carbon dioxide also allows scientists to explore one of the most complicated aspects of climate change: feedback mechanisms. I've got a feeling you are pretty familiar with feedback. You may have been in a school auditorium, let's say, when someone with a microphone on stage walks too close to one of the loudspeakers. A piercing, earsplitting sound starts coming out of the speakers. It's what we call "feedback." The moment the feedback begins, the person holding the microphone immediately tries to cover it with a hand, and runs away from the loudspeaker. That loud feedback screeching is a consequence of the sound from the loudspeaker feeding back into the microphone, which sent the electronic signal of the sound to the loudspeaker in the first place.

If the person on stage doesn't intervene by covering the microphone, or turning the amplifier off, or yanking a cable loose, that noise would get as loud as it could, and it would stay that loud until something breaks. This is called "positive feedback." Since the energy involved is turning (or returning) or looping back on itself, we call it a positive or amplifying "feedback loop." Now you're really ready to understand the mechanism of climate change.

For those of you who might insist that you are bad at mathematics, I have to point out: No you're not. You understand feedback and you know the difference between adding and subtracting. That means you have the math skills to understand climate science. Whether you realize it or not, you and I subtract all day. When you turn on the shower and stick your hand in the showering flow you instantly compare the temperature of the water with the temperature you seek, and you do that by subtracting (in your subconscious) the temperature you want from the temperature you're feeling. If the water is too cold, you subtract a lower temperature from a higher temperature and you are informed that you need to add heat: more hot water. If on the other hand—wait, it's almost certainly the same hand—the showering flow is too hot, you still, in your intuitive mind, subtract the same two temperature variables in the same order. When you subtract a bigger number from a smaller number, you get a negative number. In this case, you intuitively subtract the higher temperature from a lower one, and you get a negative value or number. That just means you want what you might call "negative hot water": you want more cold.

In climate science, a positive feedback loop describes an effect that is intensified by nature. A negative feedback loop describes an effect that is reduced or attenuated by nature. Climate change is driven by energy—here we are talking about feeding back energy—so it either adds to, or subtracts from, the overall energy in the atmosphere and ocean. We think of Earth as a climate system, and an energy input as a "forcing function." In the case of the public address microphone creating that screeching, electricity is the energy source. It amplifies sound. The electrical energy in the amplifier is forcing the sound energy to become bigger. The microphone makes the loud sound into more electrical energy, which the amplifier forces to become more sound energy, which in turn becomes more electrical energy, forcing more sound, and so on. In climate systems, we refer to the energy driving the feedbacks in shorthand; we just call them "forcings." For example, the Sun's energy is a forcing.

Earth's climate system has several forcings and feedbacks, both positive and negative. That is why simply knowing the amount of carbon dioxide in the atmosphere doesn't tell you exactly what the climate will do next, or what it did in the past. It turns out that water vapor by itself actually has an even more powerful greenhouse effect than carbon dioxide. But when it rains or snows, the water falls out of the atmosphere, whereas $CO_2$ persists.

As carbon dioxide and other greenhouse gases cause the world to warm, more water evaporates, as well. With more water vapor in the atmosphere, the world warms a little more. This is an amplifying positive feedback driving the world's temperature up a little. But once that water vapor finds its way to a certain altitude, the molecules spread out, cool off, change to tiny droplets of liquid water, and we call them a cloud. White puffy clouds reflect sunlight into space, which cools the world off a little bit. The process becomes a negative feedback.

As you might suspect, since clouds take on so many different shapes at so many different altitudes, it gets a notch more complicated. Puffy low clouds act as a negative feedback, reflecting sunlight and cooling the world. But when wispy, icy cirrus clouds form, they're very high in the sky. They reflect heat back down toward Earth's surface and produce a net positive feedback. Since water vapor in the atmosphere makes different types of clouds that behave in different and complex ways, a computer model that can accurately predict the future of our climate needs to be complex as well. But in computer science, we love complexity. That's why we have computers. All the models show our world is warming due to our activities and nature's feedback systems.

Volcanoes noticeably influence climate, but not through $CO_2$. Major eruptions can spew megatons of sulfur dioxide gas into the air. Mount Pinatubo, which erupted in 1991, is estimated to have added about 20 million tons of sulfur dioxide to the atmosphere at an altitude of about thirty kilometers (twenty miles). Up there, the sulfur dioxide mixes with water vapor and forms an aerosol of sulfuric acid droplets. These yellowish droplets reflect sunlight into space, so Earth cools off a little. This effect is especially measureable when the volcano is in the tropics, near the equator. Volcanoes therefore are mostly a negative feedback in the climate system. I say "mostly," because volcanoes also emit carbon dioxide, which warms the world in positive feedback fashion, but human activity puts about one hundred times as much $CO_2$ into the air as all of Earth's volcanoes combined. It ain't even close. And in the long run, any cooling from sulfuric acid droplets pales in comparison with the warming effects of human-generated greenhouse gases.

The classic climate-change positive feedback system is arctic ice. It's highly reflective. When it melts, it exposes seawater, which is dark. So the more the ice melts, the more sunlight the ocean absorbs. The warm ocean melts more ice. Over the past forty years, Arctic sea ice has been decreasing at a rate of about 5 percent a decade. (Don't be fooled by reports that the ice has recovered for one month or one year; it's the long-term trend that matters.) In the not-too-distant future, people operating cargo vessels will be able to exploit the opening of what navigators have for centuries called the "Northwest Passage." Ships will be able to sail, ice-free, from Eastern Canada to Siberia all year. Shipping companies and the military will be able to move cargo and equipment from Europe to Eurasia without having to avoid ice floes. They will no longer have to sail way, way to the south around the Cape of Good Hope, or Cape Horn, or work their way through the locks across the Isthmus of Panama. But the loss of this ice could also be a very troublesome thing: Without that ice, we're looking at warmer seas, sea life die-offs, and disruption of ancient ocean currents, which will lead to disruption of weather patterns everywhere. Because of the positive feedback nature of the interactions of the air, the ice, and sea, things are going to start changing faster and faster. Because of climatic forcings and feedbacks, humankind is changing the ocean, and the ocean is big enough to change everything.

I think back on how I ran that needle through Tennessee Ernie Ford's "Sixteen Tons" over and over. There was a message in the vinyl that I didn't grasp until pretty late in life. Coal mining isn't just back-breaking labor. It hasn't just made life miserable for a whole lot of miners over the last couple hundred years. Burning coal (and natural gas, and petroleum) has unwittingly caused us to change the climate of our world and set in motion feedback loops that are going to be very, very difficult to tamp down. There is already an enormous amount of carbon dioxide in our air that will be there for thousands of years to come and keep our world warmer than it would have been for thousands of years to come.

The plaintive clarinet and Ford's grim matter-of-fact performance in that song can remind us that we've taken the first step. We recognize the problem. We see what our carbon ways are doing to the atmosphere. Now we can get to work curtailing carbon production and get to work producing our energy with new, clean, and frankly exciting technologies. We've made a little progress already—just a little— by using more natural gas and expanding the role of wind and solar energy. It's time to stop shoveling that coal full-stop and start capturing energy we find in the air and the sky.

Now that it's clear what we need to do, let's take a closer look at how we need to do it.

# 6

# THERMODYNAMICS AND YOU

It seems that no matter how much time I spend straightening my desk, its clutter keeps spreading. The carefully piled papers somehow unpile. The envelopes, whether carefully slit open or hurriedly torn, accumulate with letters restuffed sideways or paper-clipped askew. It's not only true of my wooden desk's top; it's true of my computer's electronic desktop. There are folders. There are files with odd titles and in a remarkable array of formats: documents, pictures, spreadsheets, and unzipped files. There is always disorder, and it's always spreading.

I know; I know; it's me. I'm the one creating this disorder, but it sure seems like it's bigger than I am. Some natural force is at work making my work continuously spread out into disorder. In fact there really is a natural trend toward disorder. It's called entropy, and understanding it is essential to making the world a cleaner, more efficient place.

For humankind dealing with climate change, the material I discuss for the next five or six pages is of vital importance. I understand that you might think it's more complicated than you'd like to deal with

right now; if so, just skim along. But before you do, I cannot help but
express my enthusiasm for all this by once again quoting the renowned astrophysicist Sir Arthur Eddington. Although he makes ref-
erences to other discoveries in physics that might not seem relevant,
I just want you to appreciate his seriousness and his dry wit. In a
1927 lecture, he remarked:

> The law that entropy always increases—the second law
> of thermodynamics—holds, I think, the supreme position
> among the laws of Nature. If someone points out to you that
> your pet theory of the universe is in disagreement with Max-
> well's equations—then so much the worse for Maxwell's equa-
> tions. If it is found to be contradicted by observation—well,
> these experimentalists bungle things sometimes. But if your
> theory is found to be against the second law of thermodynam-
> ics I can give you no hope; there is nothing for it but to col-
> lapse in deepest humiliation.

The tendency of natural systems to spread out and become more
disordered finds its purest expression in this lovingly named "Second
Law of Thermodynamics," the tendency of heat to dissipate as everything trends toward thermal equilibrium. We all have some experience with converting mechanical work to heat. Rub your hands
together and they'll get warm. Stretch and relax a rubber band a few
dozen times, then touch it to your lips. It'll feel warm. Roll car tires
over a road, and they'll get hot. The energy of motion is converted to
the energy of heat all the time in just about everything we do. This is
one example of a fundamental idea in science: Energy can be changed
from one form to another. Chemical energy in food can be converted
into biomechanical energy and take you up a steep set of stairs.

As a consequence of the Second Law, there is no such thing as a perpetual motion machine. Put more bluntly, there is no free lunch in physics. Every device always loses some of its oomph to heat—which is a real drag, since almost every piece of modern technology ultimately runs on heat produced (at least in part) by burning fossil fuels. The omnipresent Second Law limits the efficiency of the engine in your car and the power plant that generates your electricity. It constrains all the efforts to reduce greenhouse emissions. In short, it is the foundational challenge to anyone who wants to improve the way we live without increasing the amount of energy we use.

The battle to make a better world is a battle against thermodynamics. So if we are going to solve great problems, we better know what we are up against. The Second Law is unstoppable, so let us be unstoppable, too. Let's understand it and take it into account, so that we pursue technologies that can take us to cleaner, greener ways of using and producing our energy.

We've been waging this war ever since the invention of fire, but it really came to the fore at the beginning of the Industrial Revolution, when James Watt figured out how to drive energy in the other direction. In 1781, after a few years of tinkering, Watt developed a steam engine with a spinning shaft. When you heat water enough, you get steam. With steam, you can drive a piston back and forth in a cylinder and do work. If you can convert the back-and-forth to an around-and-around, you can spin a shaft. When you get a shaft spinning, you can do all kinds of mechanical things. You can run pumps, wood-carving lathes, sewing and textile-weaving machines. Put another way, the Industrial Revolution revolved around the shafts and axles of spinning machines, almost all of which were driven by heat. The hot steam-driven engine revolutionized manufacturing by controlling the disordered nature of molecules as they are heated and cooled. It uses the Second Law to our advantage, and we've been following that path ever since.

The reason heat can be so hard to handle or induce to do something useful is mechanical in itself. Heat engines require controlling motion on the molecular level. Molecules have motion just like bouncing balls. The thing or quantity we call temperature is a measure of the moving, or "kinetic," energy of molecules. In fact, the average kinetic energy of molecules is the modern definition of temperature. The hotter the temperature, the faster molecules are moving. If you let them, they'll spread out, just like the air inside an inflated balloon if you leave it untied. That is why a teakettle can whistle: The heated steam spreads out, fleeing the kettle, and so does its energy. The tendency of materials like steam to spread out is described by the Second Law of Thermodynamics. Entropy describes how much things have spread out; you could say it describes the amount of disorder on nature's desk. It's a measureable physical property, just like speed, distance, or humidity.

It turns out that spreading out is the key to making motion. You can't get heat to do any work unless you let its energy spread out, unless you let its disorder increase. You can't get anything done—that is, you can't do useful things like spinning shafts or car wheels— unless you're willing to give up some of a system's energy to heat. In other words, you can exploit the Second Law to do useful things, but you always lose something in the process. Whether you want to make a more efficient car, a better power plant, or a superefficient way to get clean water, you need to keep this in mind. No matter what clever tricks you try, the Second Law is always a step ahead of you.

If you could shrink yourself down to the size of an atom you might have a much clearer picture of what's going on. Wait—why not? Let's shrink ourselves and look around. In any given bit of matter, be it a coffee cup, a baseball, or a bucket of water, the molecules are moving at different speeds. Some are going fast, some are going slow. When we measure the temperature, we are actually measuring the average of those molecular speeds. That's why a puddle of water can evaporate even though it is nowhere near its boiling temperature. Some of the molecules are going faster than others; those faster ones escape and become vapor. After a few molecules have left the liquid, the heat energy available to the puddle is redistributed to fewer molecules. There's more energy per molecule, so more of them evaporate, until the whole puddle is gone.

I'm picking water as an example for a good reason. Most of the electrical power in the world—everything produced from coal, oil, natural gas, or nuclear material—relies on boiling water to spin a turbine that runs a generator. Water molecules are wonderful, complicated things. They have a tendency to stick together (you can feel that as surface tension, the "skin" on top of any open bit of water). It takes quite a bit of energy to turn liquid water into a gaseous water vapor. And it becomes a limit. We can add heat and add heat, but unless the molecules have a place to go, we can only get them to do so much work.

James Watt and the brilliant English scientist William Thomson, later known as Lord Kelvin, figured out ways to extract as much useful work as possible by managing the temperature of the steam and the temperature of the area to which the steam from their engines was exhausted.

One of their insights is that there is no such thing as "cold," at least in this scientific sense. There is instead only the absence of heat. Oh, I love the time-honored joke: "A Thermos bottle keeps hot things hot and cold things cold—how does it know?" Well, okay, there's no cold, but what if there were no heat at all? None. Put an inflated balloon in a refrigerator, and it shrinks. Put it in the freezer, and it shrinks some more. It turns out that there's a theoretical, but relatively easily calculated temperature, at which the balloon would shrink to nothing. This is "absolute zero." When you measure temperatures at absolute zero, you can compute entropy. In the metric system, absolute temperatures are measured in "Kelvins." A Kelvin is just like a degree Celsius, only measured from absolute zero. $0°C = 273.1$ K. (There is a Fahrenheit equivalent. 0 degrees $F = 459.7$ "degrees Rankine." William Rankine was a successful Scottish engineer and thermodynamicist. I used Rankine in school and through the first five years or so of my engineering career. But the metric system is much easier.)

After careful thought another investigator, the French engineer and mathematician Nicolas Sadi Carnot, expressed the efficiency of any heat engine in a simple equation. Don't panic, it really is simple. It just goes like this:

$$\text{Efficiency of a heat engine} = 1 - (\text{Temp}_{cold}/\text{Temp}_{hot})$$

The message of the equation is also quite simple. The maximum and minimum temperatures in your steam engine (or power generator, or car engine, and so on) control how efficient you can get. Furthermore, the greater the difference, the more efficient the engine can be. A typical car burns gasoline at an average temperature around 1,000°C (1,273 K). On a cool day the outside temperature might be 10°C (283 K). Plugging those numbers into the equation tells you that the very best theoretical efficiency you can get is 77%.

Well, damn. You just lost 23% of the energy in your gasoline, and that's if everything else is somehow perfect. If the tires had zero rolling resistance, if the crankshaft didn't need oil at all, if the car somehow slipped through the air noiselessly and had zero aerodynamic drag, then you could get to this number that represents barely three-quarters of the energy in the heat of the gasoline. In practice, with a real car and all its losses to friction, it's more like 28% efficient. A coal-burning power plant, maybe 32% efficient. The rest of that heat energy just goes out into the world and ultimately to outer space, spread throughout the universe and lost to us forever. It's frustrating, but that is the way it is in nature.

By the way, there's another connection between engines and climate. Meteorologists and climatologists often think of a hurricane as a giant heat engine. As the sea surface warms with climate change, the temperature difference between the sea and the sky increases a little. The Carnot efficiency of this enormous atmospheric spinning system gets just a little bit higher, and cyclonic storms can become more powerful. That is why many researchers anticipate stronger storms in a warmer world. Now, back to our regular spinning devices.

Once you've got the mechanical energy of a spinning shaft, you can generate electricity. Congratulations! But you'll always be losing a little bit of the energy you put in to heat. Engineers looking for ways to do more work with less energy keep running into the Second Law of Thermodynamics. Entropy increases. Molecular energy tends to spread out. The desk keeps getting more cluttered. You can't get something for nothing.

Whether the heat comes from burning coal or nuclear fission, heat boils water and steam turns a generator . . . and that is where most of your electricity comes from today. If we want to do more with less, then, we have only two options: Squeeze every bit of efficiency out of our heat engines, or get away from heat-generated electricity entirely.

No, wait; there is a third way. It's kind of crazy, but not completely crazy; a lot of scientists and engineers have started talking about it lately. The same laws of thermodynamics that rule our machines rule our planet as a whole. Just like in a steam engine, the heat of Earth is constantly spreading. At the same time we have the greenhouse effect preventing the Earth's heat from spreading into space, and we have more $CO_2$ making the greenhouse effect stronger and stronger, our planet is trending toward equilibrium with space. And that equilibrium temperature is much, much cooler than the Earth's surface temperature, around −15°C (0°F).

What if we changed the balance? What if we helped our planet shed its heat more efficiently, using the Second Law of Thermodynamics to our advantage? In that case, maybe we could fix the whole planet all at once, rather than trying to fix all of our energy technologies. Like I said, it sounds crazy, but is it the kind of crazy that just might work? Read on.

# 7

# FIGHTING GLOBAL WARMING WITH . . . BUBBLES?

Some evenings when I step out the front door of the Nye Laboratories West (aka my California house), I look at the crescent Moon and see Earth reflected in the sky. Now much as I would like to tell you that I have some kind of superhuman vision, this is actually a very ordinary thing. You have probably seen it yourself but may not have quite realized what you were looking at. Light from the Sun bounces off the Moon to your eyes. But sunlight also bounces off Earth and reflects off the Moon and then to your eyes. The effect is most easily seen when the Moon is a thin crescent. At that time, if you were on the lunar surface, you'd see that the Earth is nearly full and particularly bright. The result is that you, here on Earth, can see the whole of our side of the Moon: the crescent bright from sunshine, the night side of the Moon ghostly lit by earthshine.

The result is sometimes called "the new moon in the old moon's arms." It is very beautiful. And it might, just maybe, show a way to avert global warming by reengineering our whole planet—a hotly debated topic (pun intended) called geoengineering.

There are a lot of geoengineering ideas floating around these days, but the one that I find most intriguing takes aim directly at earthshine. The brightness of earthshine depends on our planet's albedo, the fraction of incoming light that it reflects back to space. The word *albedo* comes from "albus," a Latin word for white (like the albumin in a fried egg). There are different ways of measuring albedo, but roughly speaking Earth's albedo is about 0.30; that means 30 percent of the sunlight that strikes Earth bounces back into space. For comparison, the Moon's albedo is only about 0.12. Earth reflects a lot more light than the Moon, because Earth has clouds and ice.

On a typical day, any face that our planet presents to an observer in space has just shy of 70 percent of its surface covered with clouds. Puffy white cumulus clouds are almost as reflective as fresh snow, up around 75 percent. All that light coming from below your eye level is what can lead to a troubling condition we call "snow blindness." That's why skiers wear goggles and sunscreen. Heavy dark stratus clouds on the other hand are less reflective, down around 45 percent. The ice at the North and South poles is pretty reflective as well, up around 60 percent. So with over five-eighths of Earth covered in 75 percent reflective clouds, plus another significant percentage covered in white ice, Earth reflects around 30 percent and soaks up about 70 percent of the solar energy that strikes it. The math adds up.

Earth's oceans are surprisingly dark. Seawater reflects only about 10 percent of the sunlight that hits it. Those ice sheets that reflect around 60 percent have a big effect on our situation, sending incoming solar energy back out into space. As the world warms, though, the ice melts and we will have less of it, at the North Pole especially. This leads to less light reflected into space, more absorption, and an even warmer world. It's a feedback loop, a positive feedback like the ones described earlier. Warming begets more warming. This business of albedo and reflection and absorption and re-radiation is what becomes our planet's energy balance.

But what if we could control the albedo of Earth? What if we could compensate for the melted ice and warming atmosphere? In the industrial world, we have a significant area of land covered by buildings with roofs. As a first cut, doing hardly anything, we can require that all the roofs of every suitable building be white, or nearly white. I'm not talking about cathedrals or classic architecture. This would be for new houses or new roofs on old houses, and especially factories and warehouses. It would not be a trivial change. We could affect over 2 percent of the U.S. On a global scale, that could be a big deal.

My friends, if one or a few of you out there can come up with a "pale pavement," you could change the world. I'm not talking about conventional cement. It's expensive, and municipalities have to really buckle down to commit to a new reinforced concrete roadway. I'm talking about a substitute for asphalt. Instead of a jet-black parking lot, with an albedo less than 10 percent, I'm imagining a nearly white parking lot, with an albedo up around 70 percent. Parking lots would not be so sweltering, cities would be cooler, the whole world would be (slightly) cooler. It's a problem not easily solved—leastways so far. But it sure seems like an enormous opportunity for a chemical-engineering entrepreneur.

Pavement is cool, or it could be. But here's an even bigger idea. At a meeting on the Google campus a few years ago, physicist Russell Seitz described a clever way to raise the albedo of the whole planet. His deceptively simple scheme was to make the world more reflective by blowing bubbles in seawater and ponds. The phenomenon that sparked (or popped) Seitz's idea was discovered several years ago by scientists who were analyzing data from Earth-observing satellites. When a major storm, like a hurricane, passes over a patch of ocean, bubbles of air get stirred up and mixed into the sea. When nature produces bubbles in a liquid you get what's called a "hydrosol." (I'm pretty sure you are familiar with the reverse: We call droplets of liquid in gas an "aerosol.") In our ocean and lake hydrosols some air also gets dissolved completely, and that's what fish breathe. The thing here is that a bubbly water-air hydrosol becomes slightly more reflective than an airless quantity of water.

If you're able-bodied, you've seen this phenomenon for yourself when you looked out at rapids in a river. We call it whitewater. Bubbles in water produce a great many surfaces to reflect light. Next time you're scuba diving, snorkeling, or floating along on the surface of a pool, observe the bubbles produced by someone exhaling below you. The bubble surfaces shine like polished silver. At sea, the same thing happens after a storm blows through. The ocean becomes foamy white. Because the shiny bubble, maritime, post-storm hydrosol phenomenon is produced by reflective skins of bubbles, which are under the surface, the phenomenon came to be called undershine. I guess the light reflected off the surface directly is "above shine," or maybe just "shine?" Hmmm . . .

The idea, then, is to produce bubbles in water to make the water more reflective on purpose. A key is that in an air-seawater hydrosol, bubbles of just the right diameter can persist in seawater for a surprisingly long time. If you've ever swum in the sea, you may be all too familiar with the feeling that you could use a freshwater shower as soon as you dry off. The dissolved minerals, salt etc., in seawater make it just a tad more viscous than pure freshwater. In the same way motes of dust can float about in a room for hours or days, microbubbles of air, on the order of a few microns (thousandths of a millimeter) wide, can, apparently, remain adrift in the sea for a week or more. The viscosity of air holds the dust in suspension. Any little swirl or current lofts them up. Air bubbles in water are buoyant, but if the hydrosol bubbles are small enough, the viscosity of the water holds them steady.

Seitz proposed producing microbubbles on purpose not to aerate ponds for plants to grow, or in fish tanks to give fish some oxygen (applications you may know about), but to reflect light. I am pretty sure the first places we ought to try this are at reservoirs. Small reservoirs like we have around Los Angeles might be good places to run some tests. But if you're thinking big, Lake Mead, the body of water impounded behind Hoover Dam, loses over 1 billion cubic meters of water (almost 4 million gallons) every year to evaporation. If we could prevent even a fraction of that water from evaporating, that would relieve some of the crushing drought in Southern California. At the same time, the bubbles would also bounce just a little more sunshine away from our planet.

It's easy to imagine taking some of the energy that the Hoover Dam power plant produces to drive some industrial-scale electric bubble pumps that would create a hydrosol of microbubbles in much the same way the jets in a hot tub spa produce bubbly streams of water under a little bit of pressure. We could even add some sort of surfactant, something to make the water thick enough to sustain microbubbles longer than it otherwise would. This would be something akin to a biodegradable soap. If it were the right organic slippery compound, it could even nourish microorganisms in the ecosystem downstream of the dam.

Reservoirs aside, at a great many heat-driven, fossil-fuel-burning power plants, there is a cooling pond. Some are so big that you can see them in pictures taken from space. The left over, not-easily-recovered, Second Law of Thermodynamics heat is dissipated in a big, open, industrial-scale pond. What if the power plant used some of its energy to pump bubbles into the surface of the pond? It might reflect a great deal of light without affecting any natural ecosystem. It's an engineering analysis worth doing, especially if we had a tax benefit or dividend for the power plant managers to keep water in their ponds and Earth's albedo a tad higher than it would otherwise be.

As I write, the City of Los Angeles has just floated 96 million "shade balls" on the surface of its reservoirs to prevent evaporation. It's another scheme to ensure the impounded liquid water remains a liquid. Plastic spheres reduce the area of the reservoir exposed to the Sun. They reduce evaporation and prevent chlorine disinfectant from breaking down. They're recyclable. Unfortunately for this discussion, they're black; they have to be in order to keep sunlight from passing through. They are addressing evaporation, not albedo reduction, but they will keep some water in the reservoirs that would have otherwise evaporated, and they'll do it cost effectively.

Speaking of cost, if the idea of reflecting light with bubbles proves effective, it has to be cost effective as well. So there is an old idea for making bubbles that's based on a discovery Charles H. Taylor made in the early 1900s in wintertime on the Ottawa River in Montréal. As water flowed over a dam, and the river was icy, small bubbles formed under the ice. Taylor was a Canadian engineer, who was strolling (I guess in bundled-up Canadian fashion) along his local river in the dead of winter. He verified reports that if he poked or pierced the bubbles, he found they were holding compressed air. The pressure of flowing water, under the right conditions, forms compressed air bubbles in icy whitewater. Taylor went on to design a remarkable installation to produce dry compressed air almost for free on the Montréal River in the town of Cobalt, Ontario.

There's a steep set of rough, rocky rapids that locals call the Ragged Chutes. Taylor designed a water and air intake that resembled a funnel fit into another funnel. The water and a great deal of entrained air fell straight down a shaft 104 meters (342 feet) to a large underground chamber at a low point in the river downstream of the Chutes. The water flowed out of the chamber through a carefully placed pipe and back into the river farther downstream. Above the pool in the chamber, the system created an enormous supply of air compressed to over 860 kilopascals (125 pounds per square inch). Furthermore, despite it being above a pool of water, the air was pretty dry. As you may know, rivers in that area are cold, which caused a good bit of the moisture to condense right out of the squeezed-up air. That part of Canada was a big mining area, so Taylor sold energy in the form of compressed air to several mining customers not far away. Compressed air is easy to transmit through regular old plumbing pipes. Miners used it to drive jackhammers, pumps, and other machinery. Brilliant.

This arrangement of falling water and pipes to produce compressed air is often called a "trompe," from the French for "trunk" or "big tube." These were used for centuries to produce compressed air for smelting or purifying iron ore to produce useful iron. Suppose we had trompes at big dams that we could use to produce hydrosols of microbubbles that we could, in turn, use to keep reservoirs in arid areas around the world from evaporating. Keep in mind that energy needed to make the compressed air would not be free. We would be using some falling water, some of the energy stored behind the dam, to produce the compressed air. It's a tradeoff between the value of the electricity and the value of the freshwater itself. Both are needed; that's why big dams get built in the first place. It also takes energy to impound freshwater, especially if we need to pipe it in or extract it from seawater. In the energetic scheme of things, adding bubbles to our reservoirs might be the right thing to do.

When I originally considered using tiny bubble hydrosols to conserve water, it sounded a little odd—well, it's not an obvious idea. But I have no trouble imagining a day when every reservoir behind every dam in warm climates, like the American West and the Middle East, would have such a system. It's an idea so crazy that, yes, it just might work. Perhaps future generations will take microbubbles in reservoirs for granted. The reason people haven't already enacted such a scheme is that, until climate change and our burgeoning human population became significant, the dams were big enough, and there always was sufficient water behind them to meet humankind's needs.

Adding bubbles to the world's reservoirs would be a major undertaking, but it's still small potatoes compared to what Seitz really has in mind. He's talking about doing the same thing to the ocean, or at least to large swaths of the ocean. Like most geoengineering ideas, it sounds crazy at first. Then I consider the dams we have already built in the western part of the United States. The water that forms Lake Mead is controlled by the U.S. Army Corps of Engineers far upstream, using water from Lake Powell behind the Glen Canyon Dam 800 kilometers (500 miles) away. We humans are already engineering our planet on enormous scales.

When I think of it this way, my conversations with Seitz make sense. In an even bigger picture, engineering the whole planet's albedo doesn't seem all that crazy after all. Here's how it might work. What if we had ships that produce the right-sized microbubbles in their hydrosol-generating wakes, and these ships would sail around right next to the arctic ice? The bubbles would reflect light and cool the seawater by the ice, which would in turn, keep the ice sheet from melting as much, which would reflect a little bit more light into space, which would slow climate change, albeit just a little.

As we carry this idea to what I imagine is its natural extent, suppose we put these bubble-makers on all of the forty thousand or so enormous cargo ships that are plying the high seas every day. They could be required to generate, or be given tariff credit for generating, huge swaths of microbubble hydrosols. The ships are out there anyway. By my rough calculations, they could probably affect about one thousand square kilometers of ocean. Okay, that's barely a 10,000th of Earth's area presented to the Sun, but 1/10,000th is still infinitely more than zero. In total, we're talking about 173 billion watts. That's not zero either. It could have a wonderful effect. After all, it has taken every industry on Earth to create this problem over the last couple of centuries. All that we can do to mitigate climate change has got to be for the good. This is especially true if we consider the economic implications. If carbon-reduction credits were included in these calculations, it sure seems like a cool (pun intended) idea.

We'd be changing a fraction of the world's overall albedo, it seems reasonable to me that the ocean ecosystems would be okay. It may even be that stirring in atmospheric gases, oxygen especially, might help some living things get along out there under the high seas. Microbubbles could turn down global warming a little, and perhaps make the oceans healthier in the process.

Who would pay for the ships and for the bubbles? Would the ships be belching greenhouse gases and undoing the whole idea? There is a lot of ice on an ice sheet. Would the adjacent sea ice idea even work? These are questions I want all of us to participate in seeking answers for. We need to consider the crazy ideas. We need to run tests. And while we are remaining open-minded about far-out ideas, there are a few other geoengineering concepts for us to consider. Honestly, I think each of these ideas is unworkable, but perhaps I don't know the whole story. Other smart people are talking about them. They are worth careful consideration, at least.

Green plants take carbon dioxide out of the air and use that carbon to make more plants. Many people are under the impression that the chemicals that become the roots, stems, and leaves of plants come from the soil. Actually most of the chemical mass comes right out of the air. A tree is a column of carbon. A forest is a collection of carbon columns. So, quite reasonably, people have pointed out that if we were to plant more trees and create more forests, what's called "afforesta-tion," the trees would pull $CO_2$ out of the air.

If they were the right kinds of trees, the carbon would be removed or sequestered for centuries, until the trees died and decomposed. By then, future generations would have time to address their version of a changed climate. It's not clear that afforesting a large area of Earth could do enough to meaningfully reduce the levels of carbon dioxide in the atmosphere, however. So researchers have proposed building resin-coated artificial trees to chemically pull $CO_2$ out of the air, day and night. They would resemble rally posters on sticks. Their resin filters would have to be rinsed with water, and the carbon wash would have to be put somewhere. As I write, such a system would cost about $600 per ton of removed $CO_2$. Right now, that's off the charts in ex-pense. But as our situation becomes more desperate, perhaps it won't seem so unreasonable. In the meantime, planting trees—afforestation—would be nothing but good. A tree is not a panacea, but real trees pulling carbon dioxide out of the atmosphere, providing shade, nur-turing biodiversity, and adding beauty can only help our situation.

A related concept would involve growing more green things in the ocean. Back when I had a grassy front lawn, I noticed that I could re-ally give the grass a jolt when I fertilized it with iron. (If you ever lit-erally "Eat Your Wheaties," you are taking in some iron. Try this. Put a high-iron breakfast cereal in a bottle of water with a magnet. Shake it up, and you'll see iron filings, little bits of iron stuck to the magnet. It's amazing.) Iron works on ocean plants as well as on land plants. Since phytoplankton in the upper few meters of the sea produce about half of the oxygen we breathe by metabolizing carbon dioxide, it's been proposed that we enhance their activity by fertilizing the sea. People have suggested spreading huge quantities of powdered iron on the sea surface to give the plankton a jolt, just like fertilizing your lawn. It's generally agreed, though, that the necessary scale of this idea is unworkable. You might effectively turn areas of the ocean into a layer of pond scum. Mainly, though, the iron would just speed things up and not remove carbon dioxide. The plants would just metabolize $CO_2$ faster, but once they died, bacterial decomposition would return that carbon to the air.

Here's another geoengineering idea: We could genetically modify crops so that they are literally a lighter color. If a wheat plant that currently reflects about 10 percent of the light that strikes it were able to reflect 20 percent instead, and if we planted huge swaths of this kind of wheat, we might ever-so-slightly increase the albedo of Earth. Biotechnologists might be able to take this on, but it is a tall order to plant enough crops with enough difference in their field's albedo to make a noticeable difference. It's just that a fallow field is already pretty reflective compared with the surface of the ocean, say, so it's not clear how much benefit there is to be had. With that said, it may be research worth pursuing. Keep in mind, farm fields get cycled; they get replanted every year. This albedo-boosting effect would have to be managed based on type of crop and time of year. Intuitively, it seems simpler to plant carbon-absorbing forests with existing plants, and let them grow indefinitely.

Another proposal is a kind of variant on the microbubble concept, only it would try to make the atmosphere more reflective instead. A fleet of ships with huge chimneys on them would circle the globe. On board, they'd have huge pumps and fans to suck up ocean water and spray sea salt into the sky. Up there, the salt would seed clouds, or make existing clouds thicker, thereby reflecting more sunlight into space. The idea is based on satellite observations of long trails of potentially reflective clouds above the smokestacks of large ships. A number of serious researchers have explored the concept. There may be problems inherent, however. First, clouds reflect heat down as well as up. It's not clear that these salt-spawned clouds would produce the intended cooling result; in fact, it's not clear what overall consequences they would have at all. The second problem, is the scale of the situation. Even an enormous ship is a tiny thing when compared with the ocean. One of the most compelling images you may ever see is the laser light show at Grand Coulee Dam in my beloved Washington State. They project a full-size image of an aircraft carrier onto the dam—that is, onto the tiny lower-right corner of the dam. The image of the ship is completely dwarfed by the dam itself. All I'm saying is that even a huge ship is not that big a deal in the atmospheric scheme of things. I am not sure our society could build enough ships and move enough seawater to have much impact globally. In similar fashion, the microbubble idea may only work for reservoirs and ponds. Nevertheless, I feel that both schemes deserve more analysis.

In the geoengineering scheme of things, the ideas that have received the most attention (in the media, at least) involve reflecting sunlight back into space before it can make it to Earth's surface. The first is to have airplanes fly around at high altitude and spray an aerosol of sulfates, the chemical chain that becomes sulfuric acid and hydrogen sulfide. Particles or droplets of this stuff at high altitude reflect sunlight. We know this because that's what happens when volcanoes erupt and spew sulfates high into the atmosphere. Some studies have been done on this. The idea as a first cut seems to make sense. But once again, how much of this stuff are you talking about carrying aloft? How often? Does it affect Earth in the same way a single shot from a tropical-latitude volcano does? Once we started down this road, it would be difficult to stop. We'd have to fly planes and spew aerosols indefinitely. It's the kind of thing that looks good on paper—or on a computer screen—but when you really start looking at how much you'd have to do and how often, it seems unworkable. But, as I often say, I'm open-minded.

Speaking of reflecting light into space, how about if we just blocked some sunlight before it got to Earth? Some of my colleagues in the aerospace industry have proposed that we launch and assemble enormous sunshades in space to do just that. Of course! We just lower the blinds to cool off the planet! Before you jump up out of your chair exclaiming, "That's it!" note well how much trouble we have launching even modest payloads into low Earth orbit. Putting much bigger things farther up and out ain't so easy. Also, there are thousands of commercial and government satellites in orbit. To stay out of their way, these global dimmers would probably need to be stationed out past the orbit of the Moon. Out there, they would orbit the Sun faster than we do, so the dimmers would probably need a solar sail or rocket engines to keep them in place. The shades would have to be spread out so no one part of the planet loses too much sunshine. Launching all the rockets to build the things would add vast amounts of new pollution. The whole concept gets wildly complex.

Each of these concepts would demand careful analysis. Each seems fraught with trouble. I'm an engineer, and I'd love to engineer our way out of this dramatic climate situation we've created. The more I look at each scheme, though, the more I've come to accept that not any one of them will be the answer. I keep coming back to my original idea: We have to reengineer our technology, energy source by energy source—the things we really can control—not try to reengineer the whole planet. Our unintentional effect on the planet is how we got into this mess in the first place. We have to learn how to overcome the Second Law of Thermodynamics and find brilliant new ways of making energy without drowning in carbon.

# 8

# TALKIN' 'BOUT ELECTRICAL ENERGY GENERATION

Let's say you were a space-faring alien. (I often think a few of you really are, especially certain politicians.) Every now and then, you swing by planet Earth to see what those Terrans are up to. For the last one hundred thousand Earth years, things looked pretty much the same. But today, one hundred years since your last visit, things are dramatically different. Earth now looks like it's on fire. There are lights beaming from the surface all over this world. Those Earthers have learned how to make their light from electricity. And that has changed this world!

Just look around you right now. How much of the environment you're sitting or standing in is powered by electricity? Chances are most of it. Even if the lights are off and you are not relying on electric heat or air-conditioning, consider the electric saws and sanders used to shape almost everything you can see. You cleaned your clothes in an electrically powered washing machine. The clothes were almost certainly created on electrically operated looms, and in most cases the fabric and dye were created or modified through chemistry, in turn enabled by electricity. Your food was processed using electricity, and electricity is what keeps it chilled and fresh. Almost all of those technologies have appeared within the past one hundred years.

If you think about it for just a minute, there is no way to maintain our standard of living without abundant electricity. There is also no defensible way we can deny these kinds of electric amenities to the billions who do not yet have them, or who are just now starting to enjoy them. The light that is spreading around the world should reach everyone in every corner of the globe. We must go forward, improving our mastery of electricity. And to do that, we need to understand, really understand, the energy that flows out of our wall sockets as if by magic. Only then can we figure out the best way to take the carbon out without turning off Earth's lights.

First of all, it's important to appreciate how much energy we use every time we throw a switch or operate any machine, whether it is a car, a computer, or paper mill that takes up sixteen city blocks. When I was a kid musing about energy and loving bicycles, I had an idea to create a bicycle-powered vacuum cleaner. While nominally not a bad idea, it would take a lot of pedal power, more than you can muster in real time, even if you, unlike me, happen to be an Olympic-class bicyclist. If you pedaled your bike for about six hours, and if you thought to store all of the energy in a battery, you'd have enough electricity to run your vacuum cleaner for all of about ten minutes. It's an example that gives us an idea of just how much we rely on a steady, abundant source of versatile electricity.

Electricity itself is fascinating. If you know that it consists of moving electrons, you are halfway to getting the whole idea, but there's a tad more to it. Electricity is actually a moving energy field, the pure energy of the cosmos, and that field travels at the speed of light. In some ways, it's even more magical when you know what it really is. When you use electricity to illuminate the room you're in, run a blender, toast some toast, dry or curl your hair, start your car, or click on your computer monitor, we don't produce electricity for the sake of having electricity. We use the energy of electricity to produce energy in another form, such as light, spin, or heat.

Every time you flick a switch, you owe a debt of gratitude to another brilliant English scientist, Michael Faraday. He conducted extensive early investigations into the workings of electricity and magnetism. After a few years of experimentation, he perfected a demonstration that goes like this: On a lab bench or table, say two meters (six feet) long, Faraday set up two coils of wire that were connected by parallel wires, like a toy train track with tunnels at each end. In the center of one of the coils on a suitably shaped block of wood, he set a magnetic needle compass. With the compass in a coil at one end of the bench, Faraday moved a bar magnet in and out of the other coil at the other end of the bench. The compass needle moved.

Faraday, along with a few of his contemporaries, knew that electricity flowing through wires created magnetism, a field around the wire, which can influence a compass needle. Other people had noticed that, too. But here's the crucial new thing that Faraday realized: This effect works the other way around. If you have a magnet moving near a wire, you get electricity in the wire. Faraday observed and carefully described the key idea. It's not that there's a magnet; it's that there is a moving magnet, a moving magnetic field. He moved a magnet in and out of a coil of wire, effectively moving the magnets near many wires at once, which in turn created a moving electrical field. Almost everything you touch and see all day owes its existence to this discovery, because this is how we generate electricity. Faraday invented the electric generator. As the alien passersby (or spaceship-flyers-by) would note, it utterly changed the world.

With a bar magnet and a coil of wire, you can get a small electrical current and produce a modest magnetic field, but how would one go about getting a magnetic field continuously? As is so often the case, when you want to do something over and over again, just run in circles. We generate electricity by spinning either a magnet or a coil of wire near or between magnets. The generator or alternator in your car is an example most of us have seen or heard tell of. The spin of the engine becomes the spin of a coil of wire in a magnetic field, which powers your headlights, radio, and so on. Every car is a generator on a small scale. Every power plant is a generator on a large scale. The overwhelming reason why humans burn so much dirty coal is to generate heat, to boil water, to spin a generator, to move a magnetic field, to generate electricity. That series of connections is what links modern technological progress directly to carbon and climate change.

Electricity came to the attention of Queen Victoria when she took the throne in 1837. Queen Victoria brought electricity to British cities. And electricity-fortified British power grew to the point that, as the saying goes, "the Sun never set(s) on the British Empire." When I reflect on this story, I cannot help but wonder what discoveries lie ahead for us living today. What fundamental trick with electromagnetism, or gravity, or the weak or strong atomic forces lies ahead for humans? Will the discovery of the Higgs boson redirect human history? What new idea in energy conversion is going to change everything?

After a few more years of competitive experimenting, Werner Siemens (founder of the huge industrial conglomerate that still bears his name) came up with a way for a coil of wire to be spun in a magnetic field. Electricity is then created or "induced" in the spinning coil. Once you get it started you can use the energy of the spinning coil to energize the magnetic field. The bigger the device, the bigger the magnetic field you can make, and then the greater the amount of electricity that can be directed out through wires in contact with the coil. This is energy conversion from one form to another at its finest. The spin of the electric generator can come from the heat of combustion in a steam engine, from falling water and a turbine in a dam, from steam produced by the heat of nuclear fission, or even from the chemical energy in a bicyclist's breakfast.

We commonly talk about the way electricity "flows" as if it were water. It's a good analogy. Electricity is like water, and wires are like garden hoses. This turns out to be another important idea as we go looking for ways to make the world cleaner and more efficient. Let's say you have some sort of nozzle on the end of a garden hose, and that nozzle is closed. When you open the valve at the wall of the house (or school or office), the hose stiffens. That's the pressure. It comes to your house either directly from a pump, or it's draining elegantly from a water tower not too far away. The higher the pressure, the more the hose will bulge or become stiff. This is true regardless of how big around the hose is. Whether it's a small piece of tubing suitable for an aquarium pump or a firefighters' hose as big around as your thigh, the more pressure, the stiffer.

Then there is the amount of water that will flow in a given amount of time—per minute, let's say. As you might imagine, the more water you want to flow and the higher the pressure at which you want it to flow, the more energy is required. If we want to pump a barrel of water from a well up, up, up to a water tower, that's going to take energy. It will take a force over the distance from the bottom of the well to the pipe opening near the top of the water tower. The faster you want to get that barrel of water up to the tower, the more energy you have to provide in a given amount of time. This is what scientists formally call "power." It's the energy used or expended per unit of time. Energy, in turn, is a force multiplied by a distance. These may seem like simple definitions, but they describe big and significant things—physical quantities.

You can think of power this way, as well: If you're going to make the trip on a hypothetical perfectly flat road from your house to the grocery store, you will need to exert some amount of force over some distance to overcome any friction you may encounter. If you do it in a car with a small engine, you'll get there and back in a certain predictable amount of time. If you do the same trip in a car with a big engine, you will be able to make the trip faster; you will be able to direct the energy to be delivered faster. And indeed, you would say the second car has more power. Power can be measured in watts, named after James Watt, the steam engine inventor.

Now back to the connection between hoses and wires (careful, you might get a shock). Water under low pressure doesn't flow through a hose or pipe as readily as water at a higher pressure. The same is true for electricity. There is an electrical analog to fluid pressure. It is voltage, which we measure in volts, which are in turn named for Alessandro Volta, an Italian investigator who made remarkable discoveries about electricity; he also invented the battery. It's difficult for us to even imagine a world without batteries, let alone live in such a world. Think of your mobile phone, your car starter, clocks, flashlights, and the important toys that need batteries. The mind boggles.

If you pick up a small battery, like one used in a flashlight or a small radio, you can hold both ends of the battery and feel no sensation other than perhaps that the ends are metal and cool to the touch. But intuitively, you know that if you touch the wires on an electrified fence, you'll get a shock. The difference is voltage, the electric pressure. Along with the pressure in a garden hose, there's the flow—the rate at which water goes through the hose. As with water flowing in rivers, we say the electricity flowing through a wire has a "current." More water running in a river means more current.

In general, a hose with a larger diameter can deliver more water than a hose with a smaller one. Furthermore, a hose under higher pressure delivers water faster than a lower pressure one. André-Marie Ampère investigated and quantified the flow of electrical current.

Among his other insights was his proposal that electricity is the flow of electrical particles; he anticipated the discovery of electrons. We measure electrical current in amperes, or amps for short. After we settled on an ampere, we gave credit where credit was due and assigned a unit to the number of electrons that flow down a wire carrying one ampere in one second. It's 6,241,509,750,000,000,000 electrons! Seriously. Every day you have quintillions of electrons doing your bidding for you. I hope that makes you feel, uh . . . powerful.

Summarizing what we have here: Volts measure electrical pressure, amps measure flow, watts measure power. Power over time equals the total amount of energy. That is why your electric utility bills you in terms of "watt-hours" (or kilowatt-hours) you have used.

Because volts and amps describe power in exactly the same way pressure and flow describe power, we can make wonderful use of the idea. We can measure and calculate pressure behind a dam, pressure in the wind, or flow in the tides. We can convert different forms of power to electrical power. If we can find ways to do it without burning fossil fuels—I mean without relying on the pressure in a steam-based or automobile engine—we will be able to make the world ever so much better for billions of people.

# 9

# STOP THE BURN—DON'T FRACK THAT GAS

While our burning of coal and gas has provided us in the developed world with the benefits of extraordinary transportation, agriculture, and information technology, it has also enabled us to cover Earth with a thin blanket of gas that is too thick for comfort. Although humankind did this unwittingly, we need to work our way off our dependence on burning ancient buried material as quickly as we can. Coal is the biggest problem here. Not only does it unleash the most carbon dioxide, it also releases all kinds of heavy metals and reactive chemicals while we're trying to squeeze out its heat. This is why there has been a global push to use natural gas instead of coal, until we can produce enough energy from other less noxious sources.

The developed world has, after all, taken timid steps to reduce its dependence on coal, but it has been suggested by a great many people that the developed world has no right to discourage or even ban the use of fossil fuels to bring that same quality of life to people in the developing world. Ironically, if we further enable the continued use of fossil fuels for any reason, we are effectively writing slow-death certificates for millions or perhaps billions of us. Instead, our goal must be to bring an even higher quality of life than we enjoy in the developed world to billions worldwide in the developing world—without the use or misuse of fossil fuels.

Since natural gas burns so much more cleanly than coal, people have suggested that natural gas be used as the "bridge" fuel to take humankind from a coal-based economy, to a methane-based economy, to an all-renewable-energy-based economy. Well, as tempting and straightforward as this natural gas bridge may seem, it is at best a very, very short-term notion. Ultimately, we have to stop burning gas and start leaving it in the ground instead.

As I write, the United States has become the world leader in natural gas extraction. Although it burns cleaner than almost any other fossil fuel, methane can leak into the air as so-called "fugitive" gas. It is contributing to global climate change in a big way. In part, the huge increase in gas use and in fugitive gas release is a result of improved extraction techniques, especially hydraulic fracturing or "fracking" of geologic formations at the bottom of oil and gas wells. A few years ago, fracking seemed like a panacea. But the reality is not so pretty.

Here's a little historical perspective. My beloved uncle Bud was a geologist. After a few years in the Army Corps of Engineers, he became an explosives salesman. He spent most of his career blowing stuff up. Even if I were objective, I would observe that he loved it. He used to wear a straw fedora, the kind of hat Frank Sinatra wore later in his career. Bud always said, "If you think a plastic hard hat is going to help you at a 'shoot,'" as he called them, "you are too freakin' close."

Bud was a storyteller, a raconteur, and apparently he—from time to time—fracked the occasional well. I have one of the industry's "torpedoes," as they're called. It's a steel tube 6 centimeters in diameter (2½ inches) and 1⅓ meters (4 feet) long, with a crude funnel soldered to the top and fitted with a bale wire handle, like a miniature bucket. I've seen sketches in blasting manuals showing how these metal tubes were loaded with sticks of dynamite and lowered to the bottom of an oil or gas well. Once at the bottom, the dynamite was set off with a jolt of electricity, and the rock at the bottom was fracked, or fractured (along with the torpedo tube). My uncle claimed that now and then, he would pour cold liquid nitroglycerin into a torpedo—and then just drop the whole thing . . . down the open well pipe. I'm not sure if such a story is true, but the torpedoes do have a funnel on top, which must have been for some purpose.

My uncle Bud and my cousins lived in farm country in Indiana for a while. My cousin Tom relates the story of my uncle using a few SOBs (Slip On Boosters for dynamite sticks) to frack a neighbor's water well. Bud assured the neighbor that the explosives would do the trick, and if they didn't, the neighbor could just take that well down to the sawmill, and the crew there would cut it up into postholes (that's a pretty good joke). The frack shoot worked. The fracked water well produced enough water for long showers and so on for many years after.

The stories aside, after an explosive went off at the bottom of a well, in general, the cracked-up rock would then yield more oil and gas. Just as the well itself was drilled straight down, this was fracking straight down. The fracturing happened right beneath the drilling crew's feet. Handling explosives was and is a dangerous business. (Perhaps that's why my uncle and his whole family were and still are drawn to it.) So well owners would only use explosives when a well stopped producing. In general, that kind of fracking did not, perhaps even could not, cause natural gas to burble up into some neighbor's drinking water a few kilometers or miles away.

But since those early days, the technology of fracking has been advanced to an extraordinary degree. In my uncle's day, straight down was pretty much the only way you could drill and the only direction in which you could frack. Now, however, drill bits have multiple rotating-grinder gearlike bits that can be steered as they drill. In just a few hundred meters or feet, a drill can change direction 30 degrees; in a couple of doglegs, the drill string can be going almost horizontally. Drill crews then pump down a fluid to fill the empty spaces in the rock formation (that's the hydraulic part). When an explosive is set off, the shock and force of the explosion knocks a whole lot of rock loose, much more than you could get back in the old days with just the pressure of the explosive's gas.

Perhaps you've been in a sandwich shop with a fidgety kid. He or she (face it, it's almost always he) starts playing with his food by poking a drinking straw straight down into the sandwich, extracting the straw, and sucking the sandwich plug out into his mouth. He'll get a little pastrami or cheese or what-have-you with each poke. But if that kid were allowed to poke the straw sideways, horizontally through the sandwich, he'd get nothing but cheese and meat every time. That is the idea behind modern well fracking.

Horizontal fracking is how the U.S. recently became the world's leader in natural gas production, but it's also why some lawmakers felt the practice had to be banned. In New York State, where I now live part-time, no one may frack a well for oil and gas. It's a law, and it was enacted on account of poorly regulated drilling. When you drill horizontally and frack a gas-bearing rock formation under or near someone else's aquifer or water-bearing rock, you are going to cause trouble. You may have seen footage of natural gas coming right out of people's kitchen sink taps. This is very dangerous and just plain irresponsible. Cutting back on fracking seems like the first step on the road away from fossil fuels. If nothing else, it will make gas less available and therefore more expensive, and less attractive to consumers. But if we replace that gas with coal, bitumen from tar sands, or oil shale, we are way, way worse off. It's time to move away from all of it.

Gas from the faucet is an extreme case. There are environmental problems with fracking, yes, but there are even bigger ones with tar sands and traditional coal—our worst enemy, when it comes to our atmosphere. Like humans, animals of all sorts don't fare too well breathing coal power plant exhaust. So one can argue that natural gas is not the worst thing in the world. Coal is.

When natural gas burns, it produces essentially clean exhaust gases. Methane is $CH_4$, a single carbon atom with four hydrogen atoms, each held by a single chemical bond. When we burn it, we combine two oxygen molecules (that's $2O_2$) with one methane molecule ($CH_4$) to produce carbon dioxide and water. Even the chemically squeamish among us can do the figuring: A single $CH_4 + 2(O_2)$ becomes $CO_2 + 2(H_2O) + HEAT$. Multiply and keep track; you'll see it all adds up. That $H_2O$ is water. It's hot, so it comes out as water vapor.

Because natural gas combustion is so much cleaner than gasoline or kerosene (diesel fuel) combustion, it is preferred, often required in many urban areas. Taxi companies get credit for using natural gas instead of gasoline gas (*sic*) in their vehicles. Since methane has just one carbon and four bonds to break and rebuild, it burns like crazy, but it is very lightweight. It's a gas instead of a liquid at atmospheric pressure, and it takes up a lot more room than gasoline, even when compressed and liquefied. Cab drivers I've spoken with complain that they have to go to special places to refuel, and they have to go there to refuel often. When I was in Hyderabad, India, all of the auto-taxis or auto rickshaws were natural gas powered, and there were plenty of places to refuel. Because of the crazy population densities and amazing number of vehicles crammed onto every street, Indian authorities decided to require clean-burning natural gas. The abundance of places to pump natural gas created another problem, though: Gas leaks are ubiquitous, and methane is a powerful greenhouse gas.

So far, we have collectively figured that if you have to burn something—because, for example, electric vehicles do not exist in the numbers required—cleaner is better. But the leaking fugitive gas is nothing but trouble. Once in the sky, unburned methane is much more powerful than carbon dioxide as a greenhouse gas. At least methane breaks down over decades-long timescales. After twenty years, a kilo of methane is 80 times as powerful as an equivalent amount of carbon dioxide would have been. After a century, it still remains 30 times as potent as an equal weight of $CO_2$ would have been. After 500 years, methane's potency is finally what scientists call "greatly reduced." But, people, we do not have 500 years to mess around with these troubling facts. We've got to get to work and capture this gas before it's allowed to leak. Or better yet, we should not unleash it in the first place. Methane gas is a very dangerous fugitive.

So, natural gas is not, absolutely not, a long-term answer to our greenhouse-warming problem. But in a lot of places around the world, it's the best we have for now. If you must frack, don't screw it up. Let's discourage leaks and fugitive gas. Capture it—don't toss it into the sky. We could tax methane on environmental grounds at a much, much higher rate than we do now. We could charge companies appropriately for each leak they allow. Because methane is so powerful as a greenhouse gas, let's phase it out as fast as we practically can, while we absolutely stop burning coal. It's everything-all-at-once time. Coal, gas, oil . . . ultimately, they all have to go. The real key for the Next Great Generation will be to build a society that doesn't need natural gas or fossil fuels of any kind at all.

# 10

# NUCLEAR ENERGY: TOO CHEAP TO METER . . . AGAIN

It was already Cold War lore when I learned about the USS *Nautilus* (*SSN-571*). It was the U.S. Navy's first nuclear-powered submarine. In 1958, she sailed under the ice and over the North Pole. It was not only a symbol of the United States' ability to compete with the Soviet Union; it was magical. I built a plastic model of the *SSN-571* that featured a representation of its nuclear reactor. It seemed to me as a kid that nuclear power was the greatest thing ever. That crew had so much electricity that they could make the surrounding $H_2O$ into $O_2$ at will (and just bubble the hydrogen overboard, or "underboard"). They could grow plants under electric lights and have their own fresh tomatoes. They could keep warm or keep cool at will. These guys on the world's first nuclear ship literally did not need to come up for air. Nuclear power sounded like the future not just for naval warfare, but for all of us in the supercompetitive, super-advanced United States—in the 1950s and 60s

Consistent with the success of the *Nautilus*, my buddies and I felt no surprise when we heard the oft-mentioned predictions: Nuclear reactors would soon deliver electricity "Too Cheap to Meter." The idea was that nuclear power would be so abundant, and cost so little, that there'd really be no need to even keep track of how much electricity residential customers use. So far that dream has eluded us, but for a lot of engineers, nuclear power looks manageably safe and promising all over again. Some feel it really is the only feasible energy source of the future. Like anyone else, I'd love to find that one idea, that one technology that could, with a little hard work, save us from ourselves. But when it comes to nuclear power, I have mixed feelings. Here's why. . . .

The primary appeal of nuclear energy is what it does not produce: carbon. Right now the two biggest zero-carbon energy sources in the United States are hydroelectric power and nuclear power. We get about 19 percent of our electricity from nuclear plants, 6 percent from dams, and about 5 percent from wind and solar. The rest is bad ole fossil-fuel burning. If we were to shut down our nuclear plants right now, we would have to burn even more fossil fuel and put the world that much more under the global-warming gun. The same is true in Germany, France, Japan, and many other countries.

Adding to the amount of hydroelectric power might seem like a good idea, but giant new dams are costly and environmentally disruptive. Solar and wind power sources are much easier to put in place, but the Sun does not shine at night and the wind is inconsistent. Right now, neither wind nor solar power can provide the "baseline," the electricity that is available twenty-four hours a day, whenever needed. (Of course if new, huge batteries or energy storage systems are developed, this situation might change.) The 24-7 availability of nuclear power has some prominent environmentalists like James Lovelock arguing that it is time to get past our fears and build a new generation of nuclear power plants. I'd hoped to get on board with the idea, but right now I have concerns that I very much hope we address.

Like a lot of guys, I have always loved going to the hardware store. I was excited as a youngster, when my family went to Hechinger's in Washington, D.C., to look at what it would take to build a fallout shelter. A few cinder blocks and a few dozen bleach bottles refilled with water, and we'd be okay. Nuclear war, along with nuclear energy, were presumed to be the common facts of life in the foreseeable future. Along with the scary but somehow manageable threat of nuclear weapons was the great promise of turning those atomic swords, hotter than the surface of the Sun, into gentle, warm, and wonderful plowshares of energy progress.

A nuclear power plant is another heat engine. Just like a coal-fired plant, it uses heat to boil water into steam that turns a generator and cranks out electricity. The big difference is that there is no chemical burning, and so no carbon dioxide, in a nuclear plant. All of the heat comes from radioactive decay. Under the right conditions, neutrons in uranium or another radioactive material, like thorium, will collide with atomic nuclei and set off a nuclear chain reaction, a reaction that releases tremendous amounts of heat. Just like in a basic old coal power plant, the heat can then be used to make steam and drive a turbine, which in turn spins an electric generator. It's just that doing all this with subatomic particles can complicate things.

Hold on, let's take a step back. Nowadays we take for granted that everything, including you and me, is made of atoms. Those atoms have an amazing secret history. All the solids, liquids, and gases you'll ever come across contain atoms created in supernovae, stars so massive (producing so much hydrogen bomb–type energy and so much star-style nuclear waste) that they soon reach a critical point and explode. Supernovae spew extraordinary amounts of their guts into the cosmic void, where their elements blend into and compress existing clouds of hydrogen gas to create a new generation of stars. Earth and everything on it (other than hydrogen from the Big Bang) is made from atoms that were built up in the nuclear reactors of stars. Amazing!

Radioactive elements are unstable atoms that spontaneously break down. They were born in supernova explosions, and now they are un-doing part of the nuclear process that created them in the first place. As they decay, these elements spontaneously release subatomic particles—pieces of themselves. The interactions of the subatomic particles release heat, and so provide us the means to transform the energy of ancient exploding stars into energy that you and I can use.

Now here comes another mind-bender. Physicists have discovered that everything that you can touch and see, all matter, is mostly empty space. It's almost unbelievable at first. Atoms consist of a relatively ex-pansive loose cloud of electrons and a tiny dense nucleus. Chemical energy comes from that loose outer area. Atomic energy comes from the inner part. That's why there is no burning or carbon involved, and why nuclear power is so much more concentrated. But that also makes atomic energy inherently messy in a different way. Within the vast empty volume of each atom there is a region of fantastic density, which has come to be called the "nucleus." The word was adopted from biology, where the center of a cell is called a "nucleus."

After years of careful investigation using magnets, electrical cur-rents, and vacuum chambers, we have come to understand that the nuclei of atoms hold particles with no electrical charge and particles with what we almost arbitrarily call positive charge. We refer to them as neutrons (neutral) and protons (positive). If the first is the customer, and the second is the bartender, the neutron asks, "Why didn't I get a check?" The proton bartender replies, "For you, there's no charge." The neutron asks again, "Are you sure?" And the proton replies, "I'm pos-itive!" See? Comedy is that simple . . .

Seriously, under the right circumstances, these particles can zip free and cause the celebrated nuclear chain reaction by colliding with others of their kind hard enough to overcome what are called just the "nuclear forces"; there's the weak nuclear force and the strong nuclear force. Protons are all positive, and they therefore would repel each other like crazy. So how can they be held in there in a big bundle? That's the strong force. Neutrons, which have no electric charge, are somehow held in there, too, by a combination of the strong and weak forces. The strong force exerts extraordinary binding force, but it does so over extraordinarily short distances, a millionth of a millionth of a millionth of a meter ($10^{-18}$ meters). It reminds me of a couple of magnets. When magnets are touching, pole-to-pole, they are hard to pull apart. But if you can pry them apart, their attractive force goes way, way down. It's the same with nuclear forces, but at strength levels that are astounding and over distances that are below what most of us can imagine.

As you may know, radioactivity is a property of certain materials in which subatomic particles spontaneously shoot out of atoms. You probably recognize the term " half-life." It's the amount of time required for half of the atoms in a given sample of radioactive material to change from one form of matter to another. We say elements "transmute" from one to another. As they do, the number of protons and neutrons changes. When heavy atoms decay, you end up with fewer protons than you started out with. The element changes from having more protons to having fewer—from a big atomic number to a smaller one. The classic case is uranium. It turns into lead . . . slowly and all on its own. At first, protons in the uranium shoot out in pairs, bound together with a pair of neutrons. Then some of the remaining particles in the uranium nucleus change into neutrons and release more energy. The energy level of the material decreases decay-step by decay-step. And in each step along the way, the particles in the nucleus release energy—heat. It's like friction, only it's between tiny, tiny bits of stuff that are held in place by tiny yet very powerful bundles of energy. That heat is what we're after in a nuclear reactor.

A fascinating thing about nature is that we can know with extraordinary precision how many atoms will transmute from one element with a certain number of protons to another element with a different number of protons, but we cannot know which individual atom will change first or second or last. We've tried and tried, but the answer, for any one atom, seems to be unknowable. Oh Bill, come on . . . you're saying that you can know when half of them will change, but you cannot know which one(s) will change? Yes, that's exactly what I'm saying.

When this property of nature was first being investigated, people (i.e. physicists) just couldn't quite get their heads around it. I still have trouble myself. There's a randomness in nature, a statistical property of matter itself that is described nowadays by what's called quantum mechanics: the system of rules that determine the behavior of atoms and fundamental particles. Erwin Schrödinger famously came up with a thought experiment in which a cat is sealed in a box, and its fate is determined by a radioactivity detector connected to a vial of poison. He pointed out that not only can you not know whether the cat is alive or dead, but in the quantum mechanical sense, the cat is truly both alive and dead—until you open the box. It's an important line of reasoning, and it's just weird. Shrödinger's cat has become the stuff of nerds' T-shirts. Early on, Albert Einstein himself was dismissive of the whole quantum mechanical idea. He snorted, "God doesn't play dice with the universe." Well, apparently the universe does play dice, after a fashion, and radioactive decay is one example of that. Nature has both randomness and remarkable predictability.

I hope every one of us appreciates that the discoveries that make nuclear reactors possible were made along the way to developing astonishingly powerful and deadly weapons—atomic bombs. While their destructive power is terrifying, the insights that humankind had in creating them have changed the world. They've changed what we know about how we fit in, our place in the cosmos, our place in space.

The trip to the hardware store for bomb shelter supplies and building plans was fascinating to me as a kid. But I had only a vague sense of what was involved. When an atomic bomb releases a huge amount of energy, it does so in a very messy (horrible devastation) kinda' way. The history of particle physics gives us perspective. Just as it's one thing to start a forest fire with a fuel-soaked rag and a match, and it's quite another to create an efficient clean-burning furnace. Nuclear energy is relatively easy to release in a big, barely controllable fashion, but has proven to be somewhat more difficult to produce cheaply and under control.

To get the nuclei of atoms in a nuclear reactor to react is not a trivial undertaking. When atoms knock apart and release their energy, we say they "fission," or split. To get them to fission, we need to have exactly the right number of neutrons and protons, and they need to be near enough to one another to react. Technically speaking, they need to be the correct isotope (from the Greek for "same kind").

What makes an element an element is the number of protons it has. What makes an isotope of an element an isotope is the number of neutrons it has. Outside of the nucleus, protons are complemented by an equal number of electrons, and the number of electrons in turn determine the element's chemical properties. By convention tracing its roots back to Ben Franklin, we think of protons as positive and electrons as negative. That number of protons is called the "atomic number," and it determines how an element will behave as a chemical. Oxygen has eight protons. Carbon has six. When they react with each other, we know how much of each element will be involved in the chemical reaction that forms the compound we call carbon dioxide.

The number of neutrons combined with the number of protons determines how an element will behave in a nuclear reaction. The combined number of protons and neutrons gives us the "atomic mass." All uranium atoms have 92 protons. Most have 146 neutrons. We identify it by its atomic mass, which is just the combined number of protons and neutrons: $92 + 146 = 238$. We call it "uranium-238"

(U-238). We write it as $^{238}_{92}U$. But every now and then, we come across a uranium atom that has only 143 neutrons ($92 + 143 = 235$), and it's called "uranium-235" (U-235). We write it as $^{235}_{92}U$. That second one has changed the world.

By all accounts, it was Nils Bohr, the guy after whom the Bohr model of the atom is now named, who figured out this business of uranium's neutrons. He was sitting in a very dignified dining area at Princeton University. He said something like "I have it," got up, and did not say another word. He marched through the snow to his office— still saying nothing to his colleague, who had followed him. Then he made three graphs on the blackboards. One showed how natural uranium behaved as a radioactive material, in which barely 0.7% of the mixture is uranium-235, and the other 99.3% is uranium-238. The second graph showed how uranium-238 would behave by itself. The third graph showed that the reason a chain reaction had not been created, and the reason it soon would be created, was that neutrons from U-235 are going just the right speed. Neutrons from U-238 are too fast. Neutrons from other radioactive elements like thorium are too slow. But the U-235's are just right. Bohr realized apparently in an instant of insight that they needed to purify or separate the U-235. This is no easy feat.

There's only about a one percent difference in the mass of these two isotopes. Through an amazing bit of chemistry, chemical engineers figured out how to separate the heavy uranium from the not-quite-so-heavy uranium, and make it into a ceramic-like material, so it can be used as nuclear fuel.

To give you a rough idea of what's involved, follow along. First you mine uranium ore, often a mixture of oxides: $UO_2$ and $UO_3$. Then you leach it with sodium chlorate or sulfuric acid to get $U_3O_8$. At this point, it's not especially radioactive. You put that stuff in an oven or kiln with hydrogen fluoride gas, then in a second kiln with just fluorine gas. Or you can wash it in hydrofluoric acid, the crazy powerful acid used to etch the soft-white frosting on the inside of lightbulbs. The goal is to combine uranium with fluorine to form the gas uranium hexafluoride $UrF_6$. In the nuclear industry, they often just call it "hex." Nothing in the energy business is entirely simple and clean, but this material is particularly nasty.

If you're a risk taker, uranium hexafluoride is for you. It reacts with water, even, or maybe especially, moisture in the air. At regular atmospheric pressure, it becomes a gas at around 125°C. So it's hot, corrosive, and radioactive all at once. But keep in mind, it's also very useful because it can be used to produce tremendous amounts of energy. It has almost magic power. Fluorine exists in nature almost always as just one isotope: fluorine-19. So when it forms a compound with uranium, the only difference in the weight of the different molecules—$UF_6$ with U-238 and $UF_6$ with U-235—is in the mass difference of the uranium. By spinning hex gas in very fast spinning centrifuges, we can drive the heavier hex gas molecules to the outside. These are the kind of centrifuges that nuclear weapons proliferation officials worry about. If a nominal enemy starts importing certain alloys or running spinning machines in secret, officials express deep concern.

With great attention to detail, engineers can separate the $^{235}UF_6^{19}$ from the $^{238}UF_6^{19}$. Then all they have to do is blast the gas with steam, which will form uranium oxide with fluorine attached, then let the newly formed particles mix with ammonia to produce ammonium di-urinate. Dry that out, and you get powdered uranium oxide at around 700°C. Then, and only then, can you generate electricity. Simple enough? Phew . . .

It's a pretty involved business. But people have been at it for seventy years. For a nuclear reactor, uranium-235 fuel is concentrated to about 5% of the mix. It's not pure explosive uranium, but it works to make heat, lots of it. Enrico Fermi constructed the first reactor at the University of Chicago, Chicago Pile 1, out of blocks of graphite, pure carbon. The graphite atoms slow down the radioactively decaying neutrons just enough to enable the chain reaction to sustain itself. It worked beautifully. The same graphite moderator idea was used in the reactors at Chernobyl. On one fateful day, technicians let the radioactive fuel core get too hot, and it exploded.

So the key idea in a useful safe reactor is getting the heat out of the radioactive heavy uranium metal core to where we can make use of it, in a controlled fashion. Most of the reactors in service today use liquid water at high pressure. Water circulates around the uranium core. Getting the heat out is not trivial. You've probably seen the tall gracefully curving towers often associated with nuclear plants. In fact, they're used at any power plant in temperate latitudes like the ones in France or in western Washington State. In eastern Washington State, on the other hand, it gets pretty hot in the summertime. So Columbia Power Station II has no curved towers. There you'll see a different arrangement of low buildings with powerful boost fans. It's all to get that heat efficiency up.

In most commercial reactors, the water acts as a neutron moderator (slower-down-erator) as well as the coolant. If there were a leak in the fuel rods, the water would get contaminated, and the whole plant could become radioactive. So there are other schemes in which one loop of circulating water transfers its heat to another loop of circulating water. Two leaks in two loops are considered very unlikely. With that said, they had a leak at the Three Mile Island plant in 1979, and it shut down the whole town, not just the reactor building.

Another popular coolant is liquid sodium, the same element that, along with chlorine, comprises table salt. In pure form, it's a metal, actually. When it's a liquid, it conducts heat very well, and doesn't corrode pipes. You have to keep it hot, or it freezes solid and can block the cooling plumbing. But keeping things hot is usually the least of the engineering challenges in a reactor. And as you may know, if sodium comes in contact with water, look out—it's explosive. It's just one more thing that makes reactors complex, one more thing to manage as reactors get more powerful and as there are more of them. The coolant, be it water or sodium, is radioactive for a brief time, but after a few hours or minutes, it gives up its radioactivity to its surroundings and reenters the reactor ready to circulate and cool again.

Getting the neutrons in the core to go just the right speed is essential to any successful reactor design. Graphite is good. Water is okay. Heavy water is even better. Heavy water carries an extra neutron on its hydrogen atoms. Such atoms are common enough. You just have to spin a huge amount of water in a centrifuge or capture the heavy $H_2O$ molecules from the circumference of a hydroelectric dam's turbine, run it through a cascade system, and let it accumulate. In any configuration, a nuclear power plant is complex. Over the years, engineers have worked with a great many different designs. You can get a sense of this from the bagful of acronyms the nuclear industry uses: PWRs (Pressurized Water Reactors, like the one found at Diablo Canyon in California), PHWRs (Pressurized Heavy Water Reactors, like the one at Point Lepreau and Rajasthan), BWRs (Boiling Water Reactors, such as the one at La Salle), and ABWRs (Advanced Boiling Water Reactors, like the one in New Taipei City). In France and her neighbors there are EPRs (European Pressurized Reactors). A more complete list would include LWRs (Light Water Reactors), LMFBRs (Liquid Metal Fast-Breeder Reactors), etc.

The complexity of all of these reactor technologies is part of what gives me pause about nuclear power. Sure, we fly around all the time in complex airplanes. We carry phones with software more complex than any one engineer could create. So, maybe it all comes down to risk. A single nuclear accident made a significant area of the Chernobyl region uninhabitable, and it may remain so for centuries or more. Then again, coal plant fine-particle exhaust kills more than twenty thousand people every year in the U.S. alone, albeit indirectly. The emissions also kill other species of wildlife in huge numbers, and the greenhouse gas emissions from coal plants are nothing but trouble. And when it comes to natural gas, it burns cleanly enough, but it still pumps out greenhouse gases. The carbon dioxide will hold in heat for centuries; the unburned fugitive natural gas is more powerful than the carbon dioxide would have been, and it lingers for decades. You can see why I'm torn.

It's also reasonable to say that the greatest obstacle to embracing nuclear power is the public perception of its danger. For example, recall the anxiety that followed the Fukushima disaster, which was a result of an earthquake and subsequent tsunami. You could ask, "Why did they put a nuclear plant right there?" Well, the earthquake was bigger than anyone thought possible. Nevertheless, there were no deaths from radioactivity. The wall of water and collapsing buildings were the problems. Four years after shutting all of its nuclear plants down, Japan has restarted a reactor in Sendai, about 90 kilometers (50 miles) from Fukushima, with new safety standards in place. There have been a great many protests in the area. Surprising as it might seem, the deaths from nuclear accidents are very few indeed. Even using the most dire extrapolations from antinuclear activists, nuclear accidents have caused far fewer deaths than coal pollution. Rational as that argument might sound, though, most of our citizens would just not go for building dozens and dozens of new nuclear plants.

Would consumer education turn that big fraction of concerned citizens around? Could the dangers of climate change convince them that uranium and thorium, not natural gas, are the real "bridge fuels" for our future? Can we run the world with a lot of help from nuclear power until other, cleaner energy sources are ready for prime time? Let's have a look at the risk and the rewards of going nuclear.

# 11

# ONE MORE REACTOR (NO, MAKE IT TWO)

Ever since I was a kid fascinated with secretly plying the world's ocean undetected indefinitely aboard a nuclear-powered, science fiction type, Captain Nemo–style, submarine, there have been a few things that I always wondered about. If a piece of uranium the size of a bowling ball can cause all the destruction I see in the pictures of nuclear weapons tests, how much do I really need to run a ship or a city? Then where does the nuclear fuel go when I'm done with it? What is it about uranium that makes it both clean and deadly at the same time? I mean, uranium and the other radioactive metals are obviously made of powerful stuff. Why aren't we just digging up the various fissionable materials and running the whole world with them?

The answer is roughly: "It's complicated." When we pump gasoline into the tanks of our cars, the gasoline is homogenous, the same from molecule to molecule. With nuclear fissionable or fissile material, such is not the case. The atoms in typical uranium fuel are not all the same. If you managed to read the previous chapter, you know that some atoms of uranium have a few fewer neutrons than others. The

$_{238}$U is a whole 'nother, nonfissionable, variety of uranium from $_{235}$U, even though they have exactly the same number of protons and behave much the same way as regular heavy-metal chemicals.

Since nuclear fission was discovered, scientists and engineers have seen a remarkable, counterintuitive opportunity. Suppose we don't beat ourselves up and go to all the trouble to separate the lighter-weight uranium from the heavier weight. If you (we) can place the not-so difficultly processed uranium-238 in the middle of the right flow or "flux" of neutrons, it transmutes to a few different isotopes of plutonium, the main one being plutonium-239. It can be used in a reactor in very much the same way as good ole uranium-235. Plutonium-239 has 94 protons and 145 neutrons. You may recall that uranium-238 has 92 protons with 143 neutrons. Since plutonium is so intensely radioactive it can be used in a reactor designed for this particular type of fuel.

So here's a cool-sounding idea. You could continuously produce plutonium fuel by "breeding" it, as is the term of art. You begin with a reactor already fissioning uranium-235. Once you got the chain reaction started, you could maintain a continuous circuit or carousel of fuel rods that would enter the reactor as U-238, transmute to Pu-239, then in turn fission and transmute down to U-238 again, along with a few other isotopes, and so on. Along the way, you would convert some of the mass of uranium to energy, just as Einstein's famous $E = mc^2$ quantifies. On paper, it sounds (or reads) just great. It would stretch out your uranium supply so you don't have to do as much mining, and it would cut down on the amount of radioactive waste at the end—all carbon free. You could even recover a huge amount of uranium from existing spent fuel rods and pellets. In France there is an active fuel-reprocessing program. The idea is amazing. It is also inherently dangerous. Handling all of the materials involved is complex, and everyone worries about terrorists or thugs stealing some of this stuff and making a very dangerous bomb.

In 1994, I had lunch with Glenn Seaborg. He was awarded a Nobel Prize for discovering and, in a sense, inventing plutonium. He told me that his contemporaries proposed calling the new element "plutinum," and giving it the atomic symbol "Pl." L would be the next letter in the name, and it would be the logical name after neptunium, number 93. It would be consistent with the order of our solar system's planets from the Sun. Over lunch, he said, that he insisted that the new element be called "plutonium." He explained, "C'mon, Bill, 'plutonium' sounds a lot cooler." Uh, yes, Glenn, er . . . Dr. Seaborg, plutonium sounds way cooler (than plutinum). He also insisted that the atomic symbol be "Pu" (like pee-yoo), because, "This stuff stinks." Plutonium, as an especially heavy, heavy metal, is fantastically poisonous. Glenn told me that if you breathe just a few micrograms, you're dead, poisoned and radioactive. Yikes. Plutonium can get you coming and going.

Plutonium has been described by Daniel Hirsch of the Committee to Bridge the Gap as the most "exquisitely dangerous material ever produced." It stays radioactive for tens of thousands of years. So if we go to dispose of it, we are claiming that we can put something somewhere that will be dangerous, at some level, for ten times longer than the Roman Empire existed. That is a long time. The Roman Empire did pretty well (in this sense). Its government lasted for almost eight centuries. As much as I believe in my country and my government, I am reluctant indeed to claim it will be here, as we know it now, in five hundred years, let alone five thousand. That right there, I hope, gives everyone pause. But compare that with the consequences of doing nothing or hardly anything about global warming caused by our penchant for pumping out greenhouse gases. Handling plutonium safely may look pretty manageable in comparison.

One compelling current idea for such a next-generation breeder reactor is that it would use thorium. Although thorium is not quite as powerful as uranium, once it gets started fissioning it could run a re-actor. Because of this lower fissioning energy, engineers claim that the system would inherently shut itself down in case of an overheat con-dition. The proposed reactors feature a mechanism that would rotate the fuel rods in and out of the hot zone in the reactor core as uranium was bred into plutonium and fissioned again. Some scientific papers on the design suggest that such a reactor could run for sixty years with-out refueling. Of course we must all ask ourselves if we know of any machine with moving parts that could run maintenance free for that long. Well there are ship hulls and airplane fuselages that old, but if those boats and planes are still in service, they have received a great deal of maintenance. What's different in a reactor environment is just that workers can't get too near mechanical parts or systems that have become radioactive. It's an intriguing claim that we could build reac-tors of such high-reliability. After all, I say to myself, I've used a fifty-year-old sewing machine. Perhaps lubrication and antijam systems could be designed to provide anticipated remote maintenance—a system of robots to maintain the robots.

And remember, nuclear waste is not just dangerous to touch; it's dangerous to be near. Surely that is part of why it mongers (*sic*) so much fear. It's out of our everyday experience. Then again, it is very danger-ous to be near a coal-burning power plant belching out mercury and polycyclic aromatic hydrocarbons. It's dangerous to be near a welding factory billowing out heavy metals. It's pretty dangerous to be near a farm fecal waste pond, or downstream from coal or heavy-metal mine tailings . . . etc. Somehow those dangers seem more familiar, how-ever. I feel that most of us don't quite understand the actual risks of radioactivity, and fail to realize that chemically stable industrial tox-ins remain toxic pretty much forever. In other words, we have to be very careful with the by-products of power generation in heat-engine power plants, no matter how those by-products are produced.

No question, radioactive waste is nasty and dangerous. With that said, people do dispose of it. In Europe it goes in old mines, places that are believed to be stable for the next hundred million years. There are salt mines or "domes" that seem to be well suited. Where I went to school in Ithaca, New York, people have quite reasonably proposed putting nuclear waste in an enormous salt dome under Cayuga Lake, where there's a salt mine seven hundred meters below the surface. The formation is expected to be stable for several million more years. It might be the ideal place to put a few semi-tractor trailer truck containers full of radioactive material someday in the next thirty years. But will curious kids wander into the salt mine two hundred years from now and contract an unexpected cancer? How serious a problem is that? It may soon be an important point to ponder. It's the kind of easily imagined scenario that mixes my feelings.

I've visited the famous (or, more accurately, infamous) Yucca Mountain site in Nevada. At one point it was supposed to become a national repository for nuclear waste in the United States. Objectively, there is just no way that place would work for waste storage. It's bored into volcanic "tuff," as it's called. The proposal was to encase the waste material in a special stainless steel called "Alloy 22." It's lovely as stainless steel goes, but I am very skeptical that any metal container could be counted on to maintain its integrity at every joint and seam over timescales of tens of thousands of years. I hearken to stalactites and stalagmites in caves. A lot of shapes can change over tens of thousands of years of drips—alloy twenty-two, schmenty-two. Er . . . I mean an exotic stainless steel capsule notwithstanding (i.e. not withstanding much of anything for that long).

Another thing about the Yucca site is that it's above the water table. You can see a stream below you when you stand in one of the parking areas. Tests were run to make sure the tunnel that engineers bored through the mountain would be watertight. It's not; it leaked.

Any radioactive material that got entrained in the leaking water would inherently flow down through the volcanic tuff and end up in a stream that ultimately flows to Las Vegas. People who live there in Nevada do not have to be full-time geologists to understand the water, and the tuff, and the time. They will have none of it. They will never allow waste to be stored in Yucca Mountain. We must let go of that idea for political reasons alone, setting the long-term death-by-cancer problem aside. On this issue, Nevadan feelings are decidedly unmixed. The project is now on indefinite hold.

I mention all this about Yucca Mountain, because for me, it's a symptom of nuclear power's big problem—people. No matter how great the technology works in principle, it has to be able to work very, very reliably in the real world. Who thought this Yucca Mountain scheme would work? Who authorized the drilling and test after test? Who kept pushing for it, when Nevadans were against it? In short, whom can we trust with nuclear waste?

That brings me back to the issue of complexity. Are nuclear reactors inherently unsafe—too complex to trust? A lot of the public now thinks so. One big challenge is that nuclear accidents are singular and dramatic, in contrast to the quiet, slow poisoning we get from coal power. The accidents at Three Mile Island and Chernobyl helped mobilize public opposition to nuclear power, and the tsunami-related disaster at Fukushima made the antinuclear movements even more strident. From both a safety and political perspective, can we have reactors near where we need the power, i.e. near big cities? I've been to Johannesburg, South Africa, where you can visit the park in which two enormous cooling towers, intended for nuclear power plants, stand to this day. They've been painted with charming colorful graffiti images. But they are not producing any electricity. Nobody there would stand for a reactor smack dab in the middle of town. Olympia, Washington, in the U.S. also has a couple of amazing cooling towers a few blocks from the state capitol building that were never put in service because of safety concerns.

These towers were designed for reactors that were the state of the art back in the 1970s. It is very reasonable to me that a different reactor design could be made safe, or rather plenty safe enough. What we want is a reactor that is immune to or tolerant of human error, both operator error and yet-to-be-recognized design errors. One example of a future reactor might be the pebble-bed style that was proposed in the 1940s and experimented with in the 1980s. The fuel pellets are ceramic balls the size of baseballs or tennis balls. They sit in a funnel-shaped "bed." They get hot, and their heat is drawn off with a circulating gas. Carbon dioxide seems to be pretty good for this (and there is no shortage of that stuff). But since a sphere got stuck in a German reactor in 1986, the idea has languished. Perhaps it's time to try this approach again with the new design informed by that lesson learned. This, I think, is the way forward: Proceed with caution, and with the wisdom of experience.

Several ideas have emerged recently for so-called Small Modular Reactors (SMRs). The idea is that these reactors would take advantage of the speed of neutrons. If the neutrons are going too slowly, no fission happens, so the machine is inherently safe in this condition. And with fuel rods or pebbles that are the right shape, no fission happens if the neutrons go too fast. It shuts itself down either way. Furthermore, nowadays everyone wants the designs to be modular. You could build most of the components elsewhere and assemble them at the site of a new power plant without the thousands of work hours required in traditional designs. You could replace components without having to take the whole thing apart. Although this sounds like an obviously good idea—like mass-produced cars in which replacing an engine filter is easy—for many years reactors weren't designed this way. The new reactors would also not be especially large. By tradition, an SMR generally refers to a power plant that would produce less than about 300 megawatts of electricity.

Through all this, uranium is still radioactive. As it's purified, it gets more dangerous. Plutonium is just dangerous no matter what you do. So is the next-generation, modern, yet-to-be created reactor going to be safe enough for all of us? Consider the following few numbers. They provide one way—I admit, just one way—of looking at things:

At any one time, there are about 800,000 oil wells in the world. Of those, about 3,100 are at sea on drilling platforms. When just one of those burst, on the British Petroleum's Deepwater Horizon drilling platform in 2010, it was largely due to negligent maintenance of the safety systems. Readers may recall that people panicked, or expressed disgust and concern, demanding immediate action by British Petroleum to stop the gusher and clean up the mess along the coast of the Gulf of Mexico in the southern United States. It took months, and the oil tar and goo is still around along the shore as I write five years after the incident. Analysis indicates that most of the oil is unaccounted for. It ended up, not on the surface, not on the shore, but in an intermediate layer of ocean water completely invisible to cameras and observers aboard ships.

Right now, there are 433 commercial nuclear reactors around the world. If you count the accidents at Three Mile Island, Chernobyl, and Fukushima as big and bad, consider what would happen if there are a number of these comparable to the number of deep-sea platforms, say 10 times what we have now, or 4,330 reactors. There will, absolutely will, be another accident. Does this mean we should abandon nuclear power altogether? Does it mean that we should just make sure—absolutely make sure—that another accident as bad as those second two examples in particular is just not possible? Is the problem just "management," writ large? If we had nuclear engineers as good as they have in France, where reactors have run for decades without incident, could we be sure we're safe? In popular terms, is the problem just Homer Simpson? We accept risk with all of our other energy sources. We didn't abandon oil after Deepwater Horizon. But with nuclear power the issues are particularly complicated and emotional.

The risks of climate change looms so large that I have to ask again: Are uranium and thorium the real bridge fuels? Should we pursue good nuclear power plant designs right away, so that we can generate carbon-free energy right now? I believe we should, but only if the plants we design just shut themselves down if we screw up. And of course, we cannot put them where leaks end up in the ocean, or where an earthquake cracks them open. It's a question of tradeoffs. Climate change has the likely outcome of prematurely killing millions. The United Nations estimated the total death toll from Chernobyl at 4,000. With completely renewable sources as the ultimate goal, can we safely pursue nuclear power now in the meantime? Will the public go for it?

As I write, nuclear engineers believe that a new generation of reactors could be built that would be safe . . . or safe enough. If this is true, it would be great. A problem here in the United States has been the secrecy long associated with nuclear power. While creating plutonium and the hydrogen bomb, a weapon that is even more powerful than the fission bomb exploded over Hiroshima, Japan, secrecy became the norm. A great deal of nuclear waste material, including dangerous solvents and chemicals, was buried in inadequate containers a few kilometers from the Columbia River in Washington and Idaho. Some of that material has leaked over the years. The tradition of secrecy has perpetuated a tradition of mistrust of the nuclear industry. I know two engineers who quit the industry because of what they each felt were inadequate safety measures for handling nuclear fuel before and after its use in a reactor.

Along with improved technology, I would want to see a new accountable, transparent style of management running a new style of reactor. Next-generation reactors might use thorium; it's not as radioactive or as energetic as uranium. But once you get it fissioning with, say, uranium-235 or plutonium-239, such a reactor could, in principle, run for a long time, and if it overheated it would shut itself down. But in general, if there were a mechanical failure, such as a leak or jam in the fuel rod or pebble mechanisms, the reactor might be very, very difficult indeed to fix. It might be too radioactive to get near. So we'd have to just leave it there until its core calmed down enough for technicians to get in and fix it. How many such disabled reactors could we afford to have around our cities or countryside? It's another thing to think about.

We can look to the situation in France, where 80 percent of the electricity comes from a few dozen nuclear power plants. The air there is clean, and the countryside in many ways much healthier than it ever was in the last century. France has never had a significant problem with any of their plants. These reactors were built in the 1970s. The French experience is worth watching because of the so-called "bathtub curve." When we make a mechanical product like an eggbeater or an elevator, typically there are a lot of failures or breakages at first, when the product initially hits the market. Then the manufacturer solves those early problems. You've probably experienced some of this yourself with software or an old computer. The product runs fine for a long time pleasing a great many users. But as the product—the eggbeater or elevator—ages or gets old it starts to have problems again. Things start to break or jam or just stop working properly. If we imagine a graph with the number of failures on the vertical axis and time along the horizontal axis, as we move left to right, the line resembles a bathtub. The graph reflects high failures at first, a period of low failures, and then a slower but steady increase in failures as we move to the right.

The French reactors are getting old. Because of citizen concern about nuclear power in general, new plants will probably not get built there, either. Instead, the old plants will probably just get their licenses renewed. Are they still reliable? Are we reaching that bathtub sloping increased-failure stage? Will they run like venerable ships? Or will they start to fail like an antique car, with problem after problem that needs expensive attention? On the other hand, if France starts to shut down its reactors, what will replace them? Will it also be a carbon-free form of electricity?

We all have to keep in mind that the world needs energy. We will soon need more energy than has ever been produced this far to date— combined. We know this because there are going to be more people alive and clamoring for energy in the coming decades than the total number of people who have ever lived on Earth since humans became humans around one hundred thousand years ago. The whole idea is staggering. Nuclear energy holds great promise. If we could, for example, safely extract the energy in our stockpiles of depleted uranium nuclear waste, we would have enough energy to run our entire continent for decades. It seems like a logical opportunity.

But just when I start to sway to that side, I come back to the disconcertingly broad range of risks connected with nuclear power, risks that are complex and highly difficult to control. The reason nuclear fuel and the leftover radioactive material is kept at such a relatively low concentration (around 5%) is the concern about terrorists getting hold of this stuff and setting off an explosion loaded with radioactive dust. Or worse yet, unleashing a real concentrated-uranium or plutonium nuclear chain reaction weapon. This stuff could be the most dangerous thing ever, anywhere.

When I visited the Columbia Generating Station on the Columbia River in Washington State, there was a lot of security. There is a fence. There are guards with real-deal machine guns. When we went out to see the silos or casks in which the radioactive waste is stored, guards patted us down like in the movies. Not to make you uncomfortable, but they even checked our crotches for some kind of weapon (or something). All this, just because we were near the casks; we were not even near the reactor. As secure as it all seems, I am pretty sure a very motivated group of terrorists could get to at least some of the nuclear material. This would be a great deal easier if they had a confederate inside, someone who went to work there for a few years before the big dastardly plot unfolded.

Maintaining airtight security around a reactor is no easy thing. There's a common problem at nuclear plants—nuclear fleas. They're not living arthropods, they're radioactive dust. It's not uncommon for a worker to get a particle in his shoe or pant cuff, say, that is radioactive enough to set off the exit detectors at a nuclear plant. After a while, technicians get tired of chasing the fleas and trying to wipe them off with alcohol swatches or what have you. So guards sometimes just let workers go home with a tiny piece of something that is just a little radioactive. Generally no harm is done. But what if a worker on the inside did this often enough that the plant guards became inured to it, and just let the fellow or gal go regularly. Such security lapses could make it possible to smuggle out something really dangerous. It's a solvable problem, but right now, as diligent as people are, things can go wrong.

If you are following along here, you are probably ambivalent, just like I am. So here's a suggestion to help us sort through this information, which is promising and frightening all at once. I feel the next step is to build yet one more nuclear plant, an absolute state-of-the-art facility with systems that will cause it to automatically shut down if there's a problem, and systems that can redundantly maintain the mechanisms required to rotate fuel rods or spheres breeding fissionable and fertile nuclear fuel. It's a huge challenge, but it may be the key to humankind's future as our energy needs go up, and our atmosphere thickens with greenhouse gases.

In other words, let's try once more to see if a next-generation plant can be built, cheaply and safely. Let's see if these new ideas really work. We have areas with infrastructure and skilled "rad" workers. For my part I am very concerned both about nuclear safety and about climate change. As the Boeing test pilot Tex Johnston remarked after executing a barrel roll with the prototype 707 jetliner, "One test is worth a thousand expert opinions." It's worth a shot. Let's try a breeder reactor that is demonstrably safe. But let's agree right now, or right then. If there's unworkable trouble, we shall apply our intellect and treasure elsewhere.

But wait—there is a bit more. All of the nuclear reactor designs we have discussed so far involve fission. What if there were a whole 'nother way to tap into the power of the atomic nucleus, one that does away with the messiness of uranium and thorium? There is actually, though it is not at all clear when or if humans will be able to tap it. The ultimate expression of the power of atoms is in the Sun. It's in the way stars produce heat and light, the process of nuclear fusion. Protons, when off on their own, repel each other like crazy. When there is enough crushing gravity, as there is in a star, protons get crushed together at such a high pressure that they fuse. The hard-to-imagine enormous force that kept them apart is exerted over the hard-to-imagine very short distance between atomic nuclei, and what we think of as hydrogen atoms become double-proton helium atoms—in an instant. And they release huge amounts of energy each time they do it. It's the way a hydrogen bomb works. We call it fusion. Protons are fused to become helium, at temperatures of millions of degrees.

In a star like the Sun, the fusion reaction is sustained and contained by the enormous gravity. On Earth, the goal for us physics buffs is to cause fusion to happen in a magnetic bottle, a virtual container that uses a huge—I mean huge—magnetic field to hold the nuclei in place. Then we would capture that heat energy somehow, and use it to produce electricity to run everything—including the magnetic field that made the whole thing possible in the first place. So far, no one has come all that close. But they are working hard at it.

Just think. How cool would it be to have self-sustaining fusion power plants all over the world? It would be energy almost for free. Hydrogen atoms are everywhere; just glance at a bottle of $H_2O$. So far though, nuclear fusion is always in the near future. In forty years, researchers repeatedly say, we'll have it figured out. That's not soon enough to deal with the immediate challenges of climate change. But, what if there's a scientific breakthrough? What if some insight arises from, say the study of dark energy in deep space, and that changes everything just as fast as the ability to generate electricity changed everything in just a few decades? One way to make sure we never figure it out would be to stop trying.

We have to pursue nuclear fusion somewhere in the world. Various big national and international agencies are sponsoring fusion research. A handful of private companies are trying to crack the problem, too. As I write, researchers at Lockheed Martin in Palmdale, California, are hinting that they may be closer than ever. In ten years they say, they'll have a system working. A secretive company called Tri Alpha Energy is also claiming big fusion progress. I'm skeptical but open-minded. It's an unknown horizon toward which we must walk steadily.

In short, we need to keep following the "everything at once" or "all of the above" approach to address climate change. While we're giving nuclear a final try, there are other carbon-free energy sources that need a lot more development. We know that for sure, because they are already starting to change the world.

# 12

# POWER OF THE SUN

Sunlight is energy. About half of it is infrared; that's heat. More than a few times, I've tried to bake muffins and scones with sunlight. In the Washington, D.C., area as a kid, and a Boy Scout, I confess I had limited success. But later in life, when I was in Southern California, with a much better understanding of how much heat hits every square meter of my planet, I pulled it off. I baked some scones, and made a rather tasty demonstration of how easy it is to harness the Sun's nuclear power right here on Earth.

I was using concentrated solar power—basically, a curved mirror that collects sunshine and turns it into a solar oven. It really works, though there is a catch. If you try it, and I hope you do, you'll find that the cooking usually takes a long time, because most solar ovens that you can rig up at home or buy commercially don't concentrate enough energy into a small enough volume to get much cooking done, and a great deal of the cooking energy is undone or lost to the air. I had to bake my scones one at a time as a result. That's the key challenge of solar energy: There's a huge amount of it, but it's spread out.

On Earth's surface, after the atmosphere has absorbed its 25 percent
or so, we get about 1,000 watts on every square meter on which sunlight falls. In a conventional oven, we have a volume about an eighth
or a tenth that big. So to be effective, or conventional ovenlike, we have
to concentrate sunlight. At home or while camping, you need a huge
mirror or set of mirrors to do that.

But if you're a power company with access to the vast open areas
in Spain, or the American West, you can pull this off. For example at
Ivanpah, California, the Solar Power Facility has 173,500 movable
mirrors that track the motion of the Sun and direct sunlight onto a
receiver tower built about 140 meters above the desert floor. Up there,
the heat creates steam that runs a conventional turbine very similar to
a steam turbine we might find at a coal- or gas-fired power plant. But
this heat is free and creates almost no greenhouse gas emissions. That
"almost" is important. The way it's run right now, engineers have to
burn natural gas every morning to get the steam system up to its operating temperature. We're not all the way there yet.

This huge plant was championed by Arnold Schwarzenegger
when he was governor of California. He is a conservative politician in
many ways, but he sees what is happening to our planet, and felt this
was a good investment in reducing California's greenhouse gas emissions. The Solar Power Facility was designed to crank out nearly 400
megawatts of electricity. I believe it will one day. As I write, this huge
plant, which takes up about 1,600 hectares (4,000 acres), is operating
at around 40 percent of what engineers thought it would be capable
of. As time goes on, perhaps the people who work there will be able to
optimize the power production. I can imagine enhancing the system
by coupling it to a not-quite-all-the-way-figured-out molten salt-storage
system. The solar-concentrator mirrors would get salt, like regular ole
table salt, hot enough to melt. That very hot liquid can be pumped
and stored underground, ready to heat up the boiler for the next day's

run. Problems arise when that salt "freezes," or solidifies, at 800°C (1,500°F). Then the plumbing gets blocked and it's tough to restart things, etc.

Right now, solar concentrators make heat to boil water and produce conventional steam, which is used to run conventional steam turbines. They're fine; we've used those kinds of turbines for over a century. But there are other types of heat engines that might be more efficient on smaller scales. Companies have invested in Stirling-style engines that use two pistons: one to let the working fluid, usually just air, expand and drive a crank, and a second piston to carry heat from one side of the machine to the other. They are very efficient. Companies around the world are proposing building thousands of parabolic mirrors to provide heat to drive Stirling engines and run generators. There are units for sale that use about 30 square meters of shiny mirrors and are said to produce 25 kilowatts of electricity. These numbers are a result of the very high efficiencies of the mirrors combined with the high efficiency of the Stirling engines. We'll see.

Wherever you might want to put these sorts of systems, you need a lot of land and it has to be in an area that pretty much always has clear weather. The potential of these concentrated solar power systems is huge. It may be in the not-too-distant future that countries that now rely on oil, like Saudi Arabia, will one day rely on enormous solar-oven power plants like the one in Ivanpah for almost all of their daytime energy needs. And note, in areas like that the middle of the day is when you use the most energy to cool the air in buildings. That's also the time that the Sun shines most intensely. Conveniently, you generate the most power at the time when people are doing the most work and running the most energy-hungry air conditioners.

Solar power works because sunlight is energy, and so is electricity. We can convert one into the other. Concentrated solar power is not the only way to do it, however. We can also convert sunlight to electricity directly, without super-high temperatures and boiling superheated steam. Direct conversion from light to electricity is what we do with a solar panel. We call them "photovoltaic": light (photo) + electricity (voltaic). This is the more familiar way of tapping solar energy, and in the long run it will probably the much more important and commonplace way, as well.

We are able to convert sunshine directly into electricity because of our understanding of quantum physics. Light can be thought of as moving in waves, or it can be thought of as moving in packets called "photons." We believe in photons and other packets of energy generically called quanta (plural of quantum) because our theory matches our experiments. Scientists have used the concept of energy quanta to create virtually every piece of electronic equipment you've ever seen, including solar panels. Since the 1870s, investigators realized that if light strikes certain metals, such as selenium, it gives off a flow of electrons—electricity. Intuitively, you might think that the brighter the light the more electrons you get. That's true. You might also think that the dimmer the light, the fewer electrons you get. And that's also true—but only up (I guess I mean down) to a point.

It's not just the brightness of the light that matters. Turns out the color matters, too. Light toward the violet end of the rainbow is more energetic than light toward the red end. There's an essential amount of energy, a tiny amount of energy, and if you're below it, you get no electron, no electricity. But if you go up and get your photons above a certain threshold: Bang, you get an electron. It turns out that energy in nature only moves in quanta. (This is the idea that won Albert Einstein a Nobel prize—not his theory of relativity.) A quantum is the smallest amount of energy there is. So when you hear someone say such and such a thing was or is a "quantum leap," he or she is actually talking about the smallest possible leap of any kind found in nature. It's an ironic use of the phrase. However, I'll grant you that the quantum leap represents an enormous step in thought. In a quantum leap, a particle is either here, or it's there—in an instant. This discovery changed the world.

When it comes to solar panels, engineers exploit another closely related property of materials. There are some materials that conduct electricity, like metals and carbon. There are other materials that do not conduct electricity at all, like silicon dioxide $SiO_2$, which you may also know as "glass." But, there is an in-between class of materials that don't conduct especially well, and they don't insulate especially well. We call them semiconductors, sensibly enough. It is with semiconductors that we are able to exploit or make use of the science of quantum physics to produce electricity with photons.

This whole thing started once again with Michael Faraday, who noticed that silver sulfide (tarnished silver) became more conductive as its temperature went up. With regular pure, or nearly pure, metals like copper, their conductivity goes down as they get warmer (a problem with power lines). The reason for the difference is that silver sulfide is not a true conductor; it is a semiconductor. About a century later, people understood the physics of photons and quanta and were able to create similar semiconducting materials on purpose. By far the most useful of these materials is silicon, in a slightly modified form.

Pure silicon does not conduct electricity. But when it contains the tiniest bit of an impurity, the mixture becomes a semiconductor. If we add a tiny bit of aluminum, indium, or phosphorus, for example, the semiconductor ends up with some extra electrons, which can flow as electricity. If we add a tiny bit of gallium or arsenic, the semiconductor ends up missing some electrons and it, too, will conduct electricity, only in the opposite direction. By long tradition, the addition of these metals to silicon is called "doping." When silicon is doped so that it ends up with extra electrons, we say it's an "n-type," for "negative type" semiconductor. If the doped silicon ends up with missing electrons, we call it a "p-type," for "positive type" semiconductor.

Doping of semiconductors is a tricky business. It's done in a temperature-controlled vacuum chamber. Then the silicon wafers are handled very carefully in clean rooms, manufacturing areas in which the air and the people who work inside are kept very free of dust. We have been able to accomplish so much with this technology that the places where these techniques were developed and perfected became the hub of the American computer industry. There's a reason that part of California is called Silicon Valley.

The missing electrons in the p-type semiconductor are called "holes". It's as though there is a subatomic-size hole where an electron would normally be. Photovoltaic solar panels are made up of layers of n-type and p-type semiconductors, the same materials that make your smartphone go. They're stacked according to how well an n-type smooches up against the next p-type, etc. The difference between what it takes to have electrons just merrily flow from atom to atom in a semiconductor, and the energy it takes to get them to jump from one specific energy level to another, is called the energy gap, or "band gap."

We layer the materials based on their band gaps. Usually, the top layer, the layer that photons first strike, is an n-type. The very bottom layer is a p-type. The sandwich ends up with two leads or wires, so we call such a semiconductor device a diode, from the Greek for "two ways," or "two paths." The doping induces electrons, which get jolted loose by photons from the Sun, to flow in just one direction. Photon energy transforms into flowing electrical energy. It's quantum physics in action, and it's amazing.

Although the energy is free, generating solar electricity takes some careful engineering. As I experienced when I was baking scones in my solar oven, the energy of sunshine is a lot more spread out than you would like. To get a lot of electricity you need a lot of solar panel, and as you might imagine, creating solar panels with large surface areas is a tricky manufacturing business. There are panels comprised of squares of silicon each of which is a single crystal or lattice of doped (monocrystalline) silicon atoms. There are panels comprised of silicon allowed to form crystals in a random (polycrystalline) fashion. These have the potential to be less expensive than the monocrystalline ones, but right now they are not as efficient.

Improving efficiency lets you use smaller solar panels (or get more electricity from the same size). I have a watch that I never wind; it's powered by light. The face itself is a solar cell. It's about 10 percent efficient. Ninety percent of the photons that strike its surface are reflected or turned to heat. The photovoltaic panels on my house are polycrystalline, and they're about 15 percent efficient. The type of panels used on spacecraft are generally multijunction monocrystalline; they're a stack or sandwich of semiconductors, and they do a lot better. They're up to 40 percent efficient.

Right now, spacecraft-style solar panels are generally considered to be prohibitively expensive for homeowners, but as this business becomes more competitive—as climate change is acknowledged by more of us as the very real threat that it is—the price of more efficient panels will come down. It's already happening, aided by large-scale production of solar panels in China. Just think what it would mean to a homeowner, or much more important a power company, if he, she, or it could use photovoltaic panels that are three or even four times as efficient as the ones currently available. Our fraction of electricity produced from sunlight could triple, if you will, overnight (see right there, that's a joke . . . ).

Materials scientists are actively looking at better ways to get photons to push electrons. Because of the nature of the energy in photons of different frequencies and how they interact with doped silicon, we reach theoretical and practical maximums (maxima) of which photon can excite the flow of which electron. In other, more normal words, we can get solar panels to work at only certain colors, or frequencies, of light. About half of the energy from the Sun hits Earth as invisible infrared rays. What if we could tune solar panels to push electrons with these longer-wavelength, lower-energy photons? It may soon be possible by making microscopic dots, "nanodots" of special metallic molecules that are just the right size to capture photons of a specific energy. Then, theoretically, you could stack layers of dots. Each layer would be tuned to capture the energy of different photons, making use of a lot more of the total sunshine.

Along with the efficiency of solar panels, we have to consider the price. If you want to replace today's coal-tainted electricity with solar electricity, you have to do it in a way that people can afford. Researchers have developed organic, plastic-based materials that are pretty good as photovoltaics. They're not as efficient as the traditional silicon style, but they are much cheaper, meaning they could be used much more widely. Other groups are developing spray-on solar cells; you could literally paint them onto a building and turn its entire exterior into a giant solar cell. Imagine cities in which every structure is its own power plant. The potential is staggering.

Solar electricity is growing rapidly, but it is still a small part of the total energy mix. Right now, the United States produces only about 0.4 percent of its electricity using photovoltaics. In Germany, which is on what most of us would not consider an especially sunny part of Earth, almost 7 percent of their electrical grid is supplied by solar power. On good days, during brief peak production hours, solar sometimes surpasses 50 percent of Germany's energy supply. Note well though, Germany still sells fossil fuels across its borders. The switch to solar is going to be a process. But no matter how cynical you may be, the potential for photovoltaics in the U.S. and around the world is enormous.

Part of the beauty of solar energy is that it can go small. Like the solar cell in my watch, or the 4 kilowatts of solar panels I have installed on my house. Homeowners around the country are doing what I did, generating their own electricity and reducing their utility bills. In some places they can actually sell power back to the utility and become active contributors to the grid. The stand-alone potential of solar power is especially important in the developing world, where many people have no access to a grid at all. With cheap photovoltaic panels, they may not need it. Solar panels can bring them the huge benefits of regular access to electricity—though to make it really work in a mainstream and reliable way, they will also need a great battery (much more on that later).

But instead of thinking small, some solar visionaries are thinking big—very, very big. There is another application of solar panels that people have been kicking around since I was in engineering school: Hypothetically, you could capture solar energy up in outer space above the clouds, above the atmosphere altogether, and beam it down, in concentrated fashion, to an antenna here on the ground. The space-based solar photovoltaics would capture the energy of sunlight on a huge scale. I've read about proposals to build solar panels that are kilometers on a side, in orbit above Earth.

It would require multiple rocket flights and some tricky assembly in the cold vacuum of space. But let's say it could be done. Then the idea is to transmit energy down as a beam of microwaves. Waves of visible light are measured in nanometers, billionths of meters. Green light in the middle of the spectrum has a wavelength right around 550 nanometers. Microwaves are measured in centimeters, hundredths of meters; roughly speaking, microwaves are a hundred thousand times longer than visible light. The microwave beam would therefore have to land on a huge antenna, say 10 or even 100 kilometers on a side.

The energy density would be manageable. People would have to avoid the antenna area, but if you wandered in it wouldn't just cook you like sitting in a microwave oven might. You may have seen the metal screen in the glass door of a typical microwave oven. That screen reflects microwaves just like a mirror does with visible light. Because microwaves are so long, by the physics of quanta, I mean quantum mechanics, microwaves can bounce off perforated metal. Or they can be captured or absorbed by metal screens with holes the size of say, chicken wire, a few centimeters across.

The big idea of space-based power is to collect solar energy above the atmosphere—where sunshine is continuous, up where we can get at the full spectrum—and put it on Earth where we can make it almost directly into electricity. It's a big engineering idea. There may be something to it. But there are some practical serious problems with the plan. First of all, it would take a lot of rocket trips to put these enormous solar panels in orbit. That would not be cheap. Just look how much trouble we have keeping space stations aloft. It takes about $3 billion a year just to keep the one International Space Station going. The Russian Mir space station is down, burned up in the atmosphere after less than fifteen years in orbit. It would not be cost effective to keep a power station in orbit for that short of a time. It would have to be boosted now and then, and that means not letting it bend or get out of alignment. In the near-zero gravity of inertial space, that just ain't so easy with something 10 kilometers on a side.

Then there is also the problem of where to place the collecting antenna. The Four Corners region of the U.S., where Utah, Colorado, Arizona, and New Mexico intersect? Maybe. Pretty much no one could go near the antenna, and planes would have to steer clear of it. I can imagine roughly a warehouse full of environmental-impact legal documents. And then if anything . . . any little thing goes wrong, you have to go to space to repair it, while it's beaming all this energy down through the atmosphere. One more thing, before I let go of space-based power: I must point out that overall sunlight is the most distributed form of energy we know. So to concentrate and redistribute it seems counterintuitive. There's an important lesson here.

The distributed nature of solar energy is a problem only if you are thinking like a utility, trying to produce all of your power in one place. But it can be a good thing if you think about making every building into its own energy source, about making whole cities into their own grid, about bringing power to the billions who are not hooked up to the grid at all. Just thinking about a space-based solar power system highlights (pun intended) that solar power's weaknesses from an old-style industrial perspective may be its strength in the Next Great Generation's point of view.

I am optimistic about the future of solar energy, in all its forms, to produce electricity for billions of us. The Sun shines the world around. Whether it's concentrated to produce very high temperatures and drive turbines or used to convert sunlight directly into electricity in photovoltaic systems, the energy is free, it's everywhere, it's carbon-free, and it's pretty much eternal. And there's also a third type of solar energy, a really interesting kind that I haven't mentioned yet: The solar energy that heats air and sends it in motion.

If you are following along closely—and I'm sure you are—you catch my drift, or my forceful breeze. I'm talking about wind power. It's a whole 'nother extremely promising path to abundant, climate-friendly electricity.

# 13

# IS THE ANSWER BLOWING
# IN THE WIND?

To this day, I love flying kites. There is something joyous about getting the wind to do work for you. And what's more romantic than sailing with a girlfriend? You can travel great distances and there's no motor sound to interfere with whatever you are talking about or whatever you are up to. Ahhh . . . Where was I? Oh, yes, energy . . . from the wind.

The wind blows because sunlight warms the world, which warms the air. That heated air gets squeezed up by cooler air nearby. Every breeze is solar fusion energy and the laws of thermodynamics all rolled into one gentle gust. If you're a warm air molecule, in the global scheme of things, you are guaranteed to find some cooler air nearby to squeeze you and send you upward like a hot air balloon, sooner or later, because Earth is turning. It is continuously presenting a different face to the Sun and a different face to the cold blackness of space. The wind energy comes to us for free, driven by the heat of the Sun combined with the primordial spin of Earth.

Wind power is like solar power in a lot of ways. As far as humankind is concerned, the wind will never stop blowing (at least, not until the Sun boils off Earth's atmosphere in about 5 billion years). Wind power produces no carbon emissions, other than what is involved in building the infrastructure to collect it. And like sunshine, wind is spread out all across the planet. If we could harness the wind on a huge scale, we could run our whole developed world several times over. Harnessing a little bit of it is straightforward. Harnessing a lot of it, on the other hand, takes some serious science.

Right now, the U.S. produces almost 4.5 percent of its electricity from the wind. That number has been growing quickly, aided by favorable tax breaks. People in the wind industry, who obviously have a big stake in this, claim that the U.S. could get 20 percent of its energy from wind by 2030. As an engineer and a guy who has driven around the Midwest, I find that number credible, even conservative, probably an underestimation. Looking at the numbers, I believe there are huge areas yet to be fitted with wind turbines, especially along our coastlines. Citizens of the U.S. would just have to decide it's worth doing.

To many people, the term wind power still brings to mind the iconic Dutch four-bladed windmill. Any number of businesses use them in their logos and advertising: Dutch bakeries, Dutch gift shops, and Dutch beer, for examples. Or you might picture the newer, metal scoop-vaned windmills used on U.S. farms. Those windmills, along with barbed wire, are said to have tamed the land of the Wild West by enabling farmers to have reliable access to subterranean flows of water. But the relevant technology here is the modern wind turbine—those giant, three-bladed generators that are increasingly popping up on the landscape in this country and around the world.

Each blade of a wind turbine interacts with the moving air much like an airplane wing. We get higher pressure on one side than on the other by tipping the blade or wing so that it has the so-called "angle of attack." The difference in pressure between the top and bottom, or front and back, provides a push or lift. Think of that great feeling of flying a kite. There's a pull from the wind; it's exerting some force. Now imagine using that force to do work, to produce useful energy. To achieve that, the force has to be exerted over a distance. So to get energy from a flying kite, at least as a first way of looking at the problem, we can imagine a reel of string.

As the kite flies downrange, or downwind, it pulls the string off the reel, and the axle of the reel is connected to an electrical generator. This rig would work, until we run out of string. The kite can fly in a steady fashion only if there is air moving past it. We cannot capture all of the energy in the wind. The same is true of a wind turbine. The blades have to let some of the wind go past. You probably have excellent intuition about this. You've probably had a kite string break, or you've accidentally let the reel of string fly out of your hands. The kite stops flying. It sails downwind for a bit and then flails to the ground (or ends up in a Charlie Brown–style tree).

You can think of it this way: On a sailboat, the sails billow out because there is slightly higher pressure on the upwind side than on the downwind side. In the same way, if the pressure on the downstream side of a windmill blade were exactly the same as the pressure on the upstream side, nothing would happen. The blades wouldn't spin. A helicopter that had a great big flat disk like a phonograph record instead of spinning blades would act like some sort of flat parachute instead of a propeller or rotor. No pressure difference between above and below or between upwind and downwind leads to no lift or no push. But just how fast does the above-and-below, the up-and-down wind have to be moving? How little difference can we get away with? This is, as we like to say in engineering, "susceptible to analysis."

In 1919, a German engineer named Albert Betz did an elegant analysis of turbines and propellers that is still an excellent guideline.

The kinetic energy of a molecule of air or water goes as the average velocity times the average momentum. Then there's the power, the rate at which energy is being used or produced. By realizing that the air on the upwind side can be thought of as being squeezed to fit through the circle that a turbine rotor spins through, Betz came up with a coefficient of performance, or power coefficient, which expresses the maximum amount of the wind's power you can theoretically grab. You, too, can show that the ideal downstream speed is one-third of the upstream speed. Then you have to go through a bit of calculus and a bit of algebra to figure the theoretical upper limits for the energy and the power. (You don't have to derive it just now.)

Betz showed that the best efficiency you can achieve is 16/27, or 59.3%. He established the use of a handy term: the Coefficient of Performance. It's often called the Betz limit: $c_{p,max} = 59.3\%$. According to this analysis, if we're talking about a spinning-blade windmill on a tower, as we now see in windy places like Texas, Denmark, California, Britain, and Iowa, we can't do any better than this. It's the law— Betz's Law.

Part of the constraint inherent in Betz's Law is a limitation on the number of blades a turbine can have. As a first cut, you might think that the more blades we have the better. Taking that idea further, logically, you can imagine a turbine with so many blades that the wind can hardly pass through. The turbine would not spin. It would act like a parachute, or just a wall. You'd get nowhere. To get to the optimal number of blades, engineers consider how much of the area the blades sweep out in a given amount of time, or per spin at a given spin rate. The idea is to catch as much wind as possible without bumping up against the Betz Limit.

The Betz Limit explains the distinctive look of modern wind turbines. Their huge turbine blades have very high tip speeds, up around 80 meters/second (300 km/hr, 200 mph). Note that these are the speeds of the tips of the huge spinning blades. The wind such a turbine is spinning through is moving only about a fifth that fast, and it's spinning at just twenty-two revolutions per minute. That's one time around every three seconds or so. I mean these are enormous and powerful machines. A turbine like this puts out up to 1.5 megawatts, and there is no $CO_2$ emitted as it turns.

To get the most efficiency with the wind available, engineers work hard to optimize the blades of a turbine. For any propeller or turbine, the blades have more twist near the center, near the hub of the circle, and less twist out at the tips. At the center the leading edge of the blade is encountering air molecules going much more slowly than the speed that the tip is running into the air. Similarly, the outside of an old style, retro phonograph record is going faster than the label. Try putting a few words on a Post-it note, and sticking it to the outer edge of a record. Compare how much harder it is to read those words than the slow-moving words on the record label itself.

It's the speed of the blades, by the way, that has caused trouble for our bats and birds. They literally can't see, or detect with sonar, the turbine tips coming at them. Because people can see the carnage of a bird or bat whacked to a splatter by a turbine blade, people have raised protests against wind turbines. To that, I have to say two things. First, the wind industry is working to get birds to see the towers and avoid them. They are also working to help bats hear the blades and fly around the towers. But second and much more important, if you're comparing the number of birds and bats killed indirectly by the coal industry, well, there's no comparison. The loss of habitat caused by mining, the pollution in the air, the acid rain that destroys their food sources . . . these overwhelm losses to turbine blades. I'm not saying it's not a problem that needs to be solved. I'm saying, keep the big picture of fossil fuels in your mind's sharp focus.

If you've ever watched a modern turboprop plane, been to an airshow where vintage aircraft come and go, or if you pilot powerful prop planes, you've seen propellers that twist their blades. When modern short-hop passenger planes land, they twist the blades enough that the propellers act like brakes, like parachutes almost, pushing backward against the air. Such propellers are said to vary their pitch, the angle at which they encounter the air. Wind turbines that are fitted high up on towers generally have heavy-duty gear trains that vary the pitch automatically to maximize their efficiency as the wind speed varies. The slower the wind, the steeper the pitch of the blades. It's another way to do a little more with a little less. Low wind speeds are generally harder to get energy out of, but variable-pitch turbine blades eke out a bit more than they would without these mechanisms.

Efficiency also explains why the towers of modern wind turbines are so tall, typically about 100 meters (300 feet). There is a poem about sailing written by John Masefield in 1902 that features this line: "All I ask is a tall ship and a star to steer her by . . ." It's a lovely turn of phrase, but a tall ship is a big thing to ask for. The reason people built ships with tall masts and a great many sails rigged up the masts was and is to take advantage of the wind high above the sea surface. Right at the sea surface, or right above the prairie in Nebraska, the wind sticks to the sea or to the ground. It forms what we call a "boundary layer." As you move up away from the surface, the wind in the boundary layer is moving slightly faster and slightly faster still. Then, suddenly, there is a height above the surface, where the tendency of the air molecules to stick together is overcome by the main flow of the wind. At this height, the wind's moving air divides into the "boundary layer" and the "mainstream."

Depending on how warm or cold it is, and on what weather front is passing through, the boundary layer above an area of flat land—an Iowa soybean field, say—can vary from about 10 meters to 30 meters (30 to 100 feet, roughly). We want the bottom of the stroke of the spinning wind turbine blade to be above that boundary layer. So as materials have gotten better and better, and the expense of putting up wind towers has gotten more and more easily justified, engineers have been designing ever-taller wind turbines. The Statue of Liberty is 93 meters tall. Typical wind turbines being designed and built today are almost one-and-half times that tall. Their blades, measured tip to tip, cover a diameter bigger than a jumbo jet measured wingtip to wingtip. They're expensive, but so are conventional fossil-fuel power plants.

Wind energy in the U.S. used to cost about 5¢/kilowatt-hour. As I write, wind energy in the U.S., on long-term contracts with utility companies, has plummeted to about 2.1¢/kilowatt-hour. The turbines are more efficient, and constructing them has become routine rather than unusual. Also right now, the price of electricity produced with fossil fuels like coal and natural gas is a little more than half that, around 1.2¢/kilowatt-hour. That is to say, wind energy is nearly twice as expensive as fossil-fuel energy, when just assessed on its own. But notice: Fossil-fuel plants pay virtually none of the cost of dumping greenhouse gases and other pollutants into the air. Instead, we all pay for it. If the cost of global warming is included, fossil fuels are the most expensive thing you can think of.

So far, we've talked about wind turbines mounted like airplane propellers on towers. These are traditionally called horizontal-axis wind turbines. They spin generators, and those generators are mounted way up off the ground. Any maintenance has to be done way up there. There is a massive gear train up there, and very large electrical cables have to make their way down to the ground or seafloor. These elements are all inherent in the design of huge, high-up, turbine hubs.

There is a whole other class of turbines that spin about an axis that's perpendicular to the ground, like the trunks of trees. The axis is vertical, so these are generally called vertical-axis wind turbines. The primary big advantage of a vertical-axis turbine is that it doesn't matter which direction the wind comes from, it gives the turbine a push. The other style, the horizontal turbines, are placed where the wind blows from almost the same direction every day. Almost all of the designs allow for a little shift left and right, in what's called the "yaw" axis. If the wind bends around to the left or right, the whole generator plus hub plus blade assembly shifts along with it, just like a conventional weathervane. It points into the wind.

The little bit of motion that is required complicates the whole mechanism on a horizontal turbine's tower, but we've got to have it for efficiency. We also have to incorporate a control damping system to prevent the generator housing from swinging back and forth like a pendulum. It's a solvable, but complexity-adding, problem. Having the wind come in off-axis leads to big losses in power production. Just imagine it coming in completely sideways, edge on, like a karate chop. You can see right away that nothing would happen. The blades would get no spin. The wind would just whistle right on by.

A vertical turbine does not have that limitation. No matter which way the wind is blowing, the vertical turbine is engaged. The Betz Limit is still there, though, no getting around it. Vertical turbines are generally not as efficient as the rotating horizontal style. Their blades spin into the wind for only half of each revolution. As a result, you don't see the vertical type around that much, at least not yet. However, vertical-axis arrangements potentially have a couple of key advantages over big propellers on high towers. They don't care (in the anthropomorphic sense) which way the wind is blowing. And their generator mechanisms can be right at ground or sea level. There are very reasonable proposals to build enormous vertical-axis turbines. These things could be three hundred meters across. They would spin at a few revolutions per minute, and their electrical generators and mechanism would be down at sea level, where they are less likely to get blown over in extreme storms, and where they'd be much easier to service. The designs that catch my eye have long graceful arms. They would look like a flat letter V, with its vertex right at sea level.

There are also proposals to mount vertical axis turbines in storm drains and sewer lines to extract energy that is currently lost. The idea here is that the generators could be mounted above and outside of the pipes. There'd be no sludge near the electrical parts. Clever, as long as there are no leaks.

Wind direction is not the only challenge; so is wind speed. Engineers are working on that, too. It may sound incredible to those who have never experienced it, but sailboats routinely go faster than the wind. When a sailboat rigged up with a regular looking triangular sail—a sloop-rigged boat—is going across the wind, it can go faster than the wind's speed. It's not magic; the sail gathers some wind energy and instead of just getting blown downwind, it directs that energy sideways, across the wind. The water (or ice, in the case of an iceboat) holds the boat's course, while its sail gathers the wind's energy. Without being able to get purchase, or holding force with the sea or ice, it wouldn't work. You'd be like that dollar bill that blew out of your hand and flew away.

Think about this, too, the tip speed of a toy pinwheel or a massive Iowa wind turbine is much, much faster than the wind. By connecting a gear train from a spinning propeller, people have, for scientific interest mostly, designed vehicles and boats that run faster than the wind. These crazy mechanisms can drive a cart or a boat faster than the speed of the wind itself. They can do it because they can get a resisting force from the medium they're riding over—the sea, the ice, or the ground. In the case of a cart, the vehicle's gear train drives wheels that get a grip on the ground. I lived in Seattle for quite a while. Now and then you'd see a hobbyist (an eccentric sailor) rig up a boat with a propeller on top to catch the wind, and a propeller below in the hull to spin in the sea. Big fun. With a turbine propeller aboveboard and a propeller below the waterline, you can drive one of these boats straight into the wind, something that is impossible to do with a conventional sailboat.

People have proposed exploiting this faster-than-the-wind idea to run turbine blades back and forth across a prevailing wind in a figure-eight pattern reminiscent of the double-line kites that enthusiasts swing in wide arcs holding two kite strings, one in each hand. A great deal of energy could theoretically be extracted because the tips of these blades could be going very fast relative to the wind driving them. It's another one of those ideas that is worth exploring in the pursuit of abundant, clean energy. We'll see if something like that could be proven workable and profitable.

Harnessing the wind is straightforward enough from an engineering standpoint. But wind energy is in the same sailboat as sunlight. You can't rely on it to be there always when you need it; wind especially doesn't blow at the midday peak of industry's requirements. Sufficient wind to build a turbine is not available everywhere. And we do not yet have the practical means to store city-scale amounts of energy, except at hydroelectric dams. As a result, wind turbines here in the U.S. are placed mostly on the Great Plains, in areas that are often quite remote from cities. That adds a problem on top of energy storage: energy transmission, getting that wind electricity from out there to in here. Right now, the wind-power industry seems to be converging on horizontal-axis turbines with enormous blades spinning on very high towers. Nevertheless, I would be surprised if this style of turbine ends up being used everywhere. The different configurations of turbines will probably find their way to the different applications for which they are best suited. I love that vertical-slow-spinning-turbine-at-sea idea.

A limitation of building turbines that are high and huge is that they have to be on large open tracts of land, where the wind blows pretty steadily. People have proposed focusing wind through a large area and directing it through a duct at a turbine set up for this higher-speed air. The trouble, though, is that you can't get something for nothing. Capturing a large volume of air and directing it into a turbine is less efficient than just building a bigger turbine without the duct. It's another consequence of the Betz Limit. With that said, we have already, largely accidentally, built enormous ducts that cannot be moved. I'm talking about the skyscrapers and large buildings in our cities. A portion of air that is forced between buildings could be harnessed to produce electricity. The total amount of electricity you could make this way would not be all that great, but it would be one more way to make urban areas into hubs of clean energy. This approach might work especially well if the turbines were also built to look good, to serve as public art as well as to contribute a bit of energy to the local grid.

There is a proposal for a large sculpture and power-generating structure along these lines in the Dutch city of Rotterdam. The proposal calls for an enormous 200-meter-high ring through which wind would blow steadily, as it does along the Dutch coast. A spray of water droplets would be introduced into the flow, and each of these droplets would carry an electric charge. The push of the wind against these droplets in a nominally static electric field would, according to the designers, produce a usable electric current, which would in turn power the whole operation, hotel, scenic observation wheel with moving gondolas, etc. The energy would come mostly from the wind, and the machine's power would be supplemented by solar photovoltaic panels incorporated into the structure. The idea for this "Dutch Windwheel" is, well, wild. I'm not clear on how the energy would balance. It may prove to be a net loss, but still use less power than it would have without the static electrical-power production. It might serve mainly as a showcase for the potential of wind power. If it helps inspire other, more function-oriented wind projects, that is all for the good.

While I'm thinking about unconventional applications of wind turbines, I should mention that there is a completely different style of turbine that is designed to operate underwater, generating electricity from river or ocean currents. Giant underwater propellers or turbines could be anchored in certain deep rivers that have a strong flow. It would be the same kind of energy we get from a hydroelectric setup, but it potentially might leave the river largely intact. Because of the Betz Limit and loss around open turbine blades, the system would almost certainly not be as effective as turbines built into ducts or "penstocks" in a big dam. But it might work with relatively small elevation changes, and its impact on the environment might be very small. It's an intriguing idea.

In France there is a tidal energy system called a "barrage" that was built back in 1966. As the tide comes in and out of La Rance River, the energy of both the incoming or flooding tide is captured and so is the energy of the receding water as the tide goes back out. There are tunnels or ducts with turbines built in that spin to produce electricity for about eight hours at a stretch. Unlike the winds, the tides are fully predictable and can provide energy at known times every day. Just as at other dams, there has been silting and the ecosystem has changed a bit over time. But there are no greenhouse gases emitted and no waste to dispose of. It's a special system installed at a special place, but perhaps other suitable barrage spots are out there; the artificially low cost of fossil fuels has kept us from pursuing these projects, i.e. so far.

The wind, the rivers, the ocean currents, and the tides all provide choices for engineers, city planners, and visionaries who are aiding the push toward carbon-free, renewable energy. There is enough energy in Earth's wind alone to power a developed-world lifestyle for the billions who do not yet have it. We just need to find ways to store the energy that wind systems produce so that it's available when and where we need it, whether the wind is blowing or not. We need to come up with ways to put energy in the bank, so to speak.

# 14

# DOWN TO THE WIRE

Having energy to do useful work is one thing; getting it where you need it, or want to use it, is another. Moving energy around is commonplace enough; we pipe and ship petroleum just about everywhere, across entire continents. But that kind of approach moves energy at the speed of trains and trucks. When it comes to electricity, things are a bit different. Electricity moves through wires at nearly the speed of light, but it loses its potency as it goes. It would be as if every pipeline and tanker had a leak in it. To make a more efficient world, we need to find a way to plug (pun intended) the losses. The issue is especially acute for solar and wind energy, which tend to be most abundant far from the places where electricity is needed most. City buildings block wind and sunlight, for example.

I got a real feel for just how much traveling our electricity does during a restless weekend many years ago, when I was younger and wilder. I got it in my head to take a long motorcycle ride. As a young man, I figured I could manage the risk, and I sought adventure. So I saddled up my small bike and rode all the way to Grand Coulee Dam on the mighty Columbia River. It produces about three times as much electricity as the better-known Hoover Dam. Grand Coulee cranks out about 230 megawatt-hours a year. It can peak at over 600 megawatts. I lived in Seattle, Washington, for twenty-six years. I was fascinated with my electricity bill from Seattle City Light. The city of Seattle gets almost 90 percent of its power from hydroelectric dams on nearby rivers, the Skagit and Pend Oreille rivers mostly.

I started thinking about how far that electricity had to travel to get to where I was using it. It's hundreds of kilometers or miles. I rode to the Grand Coulee and spent the night in the town of Electron. It is spectacular. I realized that if I were to ride or drive that far every day to pick up my energy (in some yet-to-be-figured-out form) and bring it to my apartment, well, it would be impossible. The electrons were doing it every millisecond.

Electrical engineers work hard to carefully design power plants and transmission systems to hold onto as many watts as possible. But there are some scientific limitations. The more we know and understand the limits, the better we can do in improving our electrical system writ large: the electrical grid. It is all around us, from the tall towers strung with wires, to the conduits that carry wires along walls, to the appliances and electronics in your home. Electricity in one form or another finds its way into just about every aspect of our lives. Making our grid more efficient will take a lot of work, but our far-reaching power lines and distribution networks present us with enormous opportunities for higher efficiency just about everywhere the wires are run.

Consider the energy you might need to paint your house. Whether by brush, roll, or spray, it will take energy to get the paint on the walls. Let's say it's your energy and your brush doing the stroking, and let's add some details to the story. You're painting the outside of the house, and logically you want to start from the top and work your way down. Any dripping drops will flow down the walls and can be smoothed by your skilled hand as you work. But there's another fundamental feature of physics in this action. In order to paint the upper parts of the wall, you have to get the bucket of paint up there, and you have to get you up there. Then you have to do the second-story work. In other words, you have to transmit, or carry, some energy to a higher place before you can use it there. This example illustrates an important law of nature: It takes energy to transmit energy. This is true of petroleum portage, a paint bucket on a ladder, or transmitting an electromagnetic field along a system of power lines.

Now consider another example: water flowing through a pipe. It loses some of its energy to friction. It flows along the rough walls. Its momentum is redirected around corners and through the sharp edges of valves. That lost energy becomes heat. In the analogous case of electricity, the steady transmission of energy through conductors like wires and electronics leads to losses as well. Hold on to the charger for your mobile phone, and you'll feel some of that lost heat. Put your hand near an old incandescent-style lightbulb, and you'll feel plenty of heat. In that case, there's so much lost energy that you can bake little cakes with it. Hence the enormous success of the Easy Bake Oven.

When it's flowing just one way, we say the electricity is flowing in a direct current, or DC. Of course, in order to flow at all, it has to run in a circuit (same root as the word *circle*). Nevertheless, in a DC circuit, the electrical current is going only one way, like the sweep hand on a clock; it never runs backward. This is the sort of circuit you find in your flashlight and battery-powered toys. Just like water flowing in a pipe, electricity in wires loses energy to heat. To transmit more electric power rather than hydraulic power, instead of larger pipe, we need a larger wire. Instead of a higher pressure, we need a higher voltage. But when it comes to electricity, there is a whole other opportunity that is inherent in electricity. It goes both to and fro; it can alternate. This is the essence of alternating current, or AC.

These two processes of the flow of electrical energy have led to innumerable puns and usages of the terms AC and DC. We have rock bands and descriptions of sexual preferences based on the idea that the same energy can be transmitted in two different ways.

You experience the accumulation and movement of electricity every day. It's an experience available on command. Rub a balloon on a sweater, your hair, or somebody else's hair. Now hold the back of your hand near the balloon. You'll feel an electrical field almost like an invisible glow. You sense it because your hair follicles line up a little. In the right light, you can see your hairs line up with the field. It happens because there is a difference in electric charges set up between the surface of the balloon and your skin, and that difference is static; it stays there. That's why we use the term "static electricity."

This leads us to the idea of electricity flowing between two metal plates that are not touching each other, by means of an electric field. Imagine metal plates separated by an insulator. The easiest one for me to imagine is air. Then imagine that these two plates are hooked up to a battery. A charge will build up on the plates, one positive, and one negative. The charges, along with the plates, will sit there indefinitely. If we were to connect a big enough battery that carried enough voltage, the energy would strip electrons off the atoms in the air and make a spark. Other than that, the direct current from the battery would not flow. This is a way to store electricity until it's needed.

Now imagine these plates connected to an electric field that is changing, that is varying like the crest and trough of a wave. You can imagine that complementary charges would build up on the two plates. While the crest of the electromagnetic wave is on one of the plates, the trough of the wave could be on the other plate. Keep in mind that these waves are a way of understanding an invisible energy field. It turns out that with the building and collapsing of electric fields in this arrangement, energy can jump invisibly across the insulation. By long tradition, this phenomenon is called "capacitance," and the arrangement of plate-insulator-plate is called a "capacitor." The insulator generally takes on the old name "dielectric" ("opposite of electric").

The array of capacitors in use in our world is absolutely astounding. There are millions of them in a computer, millions of them in a car, perhaps tens of thousands in a battery-powered watch. A capacitor is one of the fundamental devices in electronics. At first, it seems crazy. Electricity moves by but it creates no sparks, just an oscillating electric field.

We exploit and control electromagnetic fields when we transmit electricity. In a power line, it's the same feature of nature with which Michael Faraday—Mr. Electricity himself—dealt. The moving magnetic field produces a moving electrical field, which in turn creates a moving field in the power line, and that field interacts with the entire planet. Here is one of the places where things don't go as well as they perfectly might. There is an exchange of energy between the power line and the ground as the electric energy moves along the line. We lose energy by having some electricity build up or get stored, in a sense, unintentionally between the power line and the earth below. It's like the back of your hand and the static field, except our electrical transmission grid involves millions of watts distributed over tens of thousands of kilometers or miles rather than just a few quintillion electrons distributed over the surface of a rubber balloon or the back of your hand.

There's another area where inefficiency creeps in. We can't just connect any old electrical appliance to just any old power line. The voltages have to be compatible. You've probably seen some big sparks when the voltages don't match well enough. We use transformers with coils or turns of wire wrapped on iron cores covered with durable insulation to step-down the very, very high transmission-line voltage to the relatively low voltage you use in your house or apartment. Transformers convert the flow of alternating-current electrical fields into magnetism. Then a moment later, they convert the magnetism to another flow of alternating current electricity on the other side, their other leg, and we can supply power at the right voltage to devices that need it. It's a marvelous little technology trick, but there are heat losses here, too. You still can't get something for nothing.

There is a speed limit for electrical transmission, and that is yet another way nature exacts its tax on our power. We create a moving magnetic field, which induces an electric field that travels at nearly the speed of light in the transmission line. When it gets to the other end, where it's being used, it gives up some of its energy to the motor pump, television, or toaster. But a circuit being what it is, part of the energy of the electric field is driven back up (or down) the power line to the generating station at the hydroelectric dam, coal-fired plant, or nuclear reactor. All of this takes time. In that time, as it gives up energy, the electromagnetic field changes in a subtle but absolutely critical way. The crest and trough of a wave of magnetism changes its position in time with respect to the trough and crest of the electric component of the wave. And therein lies trouble.

In many ways, an electromagnetic field moves through wires the same way that the energy of motion is stored and released in a mechanical spring. When the power company sends electricity to your blow dryer, some of the energy is turned immediately into heat. Some of the energy is used to drive an electric motor and the fan. The coils of wire in the motor are very much akin to the coil on the end of Faraday's lab bench. While they are inducing a magnetic field to form, they are also storing and releasing energy as the field forms and collapses with each spin of the motor shaft. That stored and released energy shoots back up the wire to the power company. For a great many years, power companies were losing energy to the coils of motors and magnetic relay switches, and they didn't quite bother to know why.

It's a remarkable example of how a great many drips can fill a bucket. With each magnetic switch or motor coil losing a little bit of energy to heat as the magnetic fields built up and dissipated everywhere in a city, state, or vast combination of states, like New York, New Jersey, Connecticut, and New Hampshire, the power companies were unwittingly heating the world (a little) and leaking away valuable electricity. Eventually, they figured out ways to synchronize most of our electrical devices so that they're "in phase." That's why modern electrical plugs have a wide prong and a narrow prong. With everyone everywhere orienting the phase of the electricity they each buy in just the same way, power companies lose far less energy, as the field bounces out and back to us users like a spronging springing spring.

We currently (pun intended) lose about 6 percent of our power just during transmission. That may not sound like much until you think what it's 6 percent of. It amounts to about 2 trillion kilowatt-hours, enough power to run another New York City or another state of Montana all day every day.

It's easy to imagine a future in which every electrical device we use in every household, factory, and farm would be synchronized to the power grid with happy, compact, built-in electronic circuits. These circuits would add a slight cost to the purchase price of any appliance or industrial machine, but in the long run, such systems would allow us to do a bit more with a bit less, and in the long run they'd cost a bit less, too, because of the energy savings. They need not be limited to wealthy countries. They could be a part of the normal way technology works everywhere there is a grid.

As developing nations electrify their landscapes, entrepreneurs could work together with engineers and regulators to reduce electricity losses in the local grid. It would be a small but important step toward creating a more efficient world. But now that we understand the nature of the problem, you are ready for me to hit you with some much bigger, bolder ideas.

# 15

# LET'S TRANSFORM THE GRID

I remember as a kid the time I thoughtfully tested what it would feel like to touch both prongs of a vacuum cleaner plug when it was pushed only about halfway into a wall outlet. It was a small-scale disaster. I got a jolt that went up my arm in an instant. I had no idea what had happened. Is the plug alive? Did it just bite my arm? Yikes! An invisible source of crazy, creepy pain! Scary stuff. I remember my father proudly pointing out the value of this experience with 110-volt electricity. It was over fifty years ago, but I remember the jolt all too well. I also remember thinking that your bones do not actually light up the way they do in some cartoons.

Over the years, I experimented. I tried getting shocks from my older brother's electric trains. I stuck my tongue on 9-volt batteries. Each of these experiences gave me a visceral appreciation for the invisible, almost magical power of electricity. I wondered: How is this done—shock versus no shock? How is electricity sent or transmitted from one place to another? The answer turns out to be vitally important in our search for a more efficient electrical grid, here in the United

States and around the world. Just look around your house or your apartment. Electricity is flowing everywhere! And it is driven by the energy in electric fields, invisible waves in the wires and the air.

We can start with wall chargers or power packs. You probably own at least half a dozen. Some of you may own four times that many. I imagine you've also noticed that no matter what you do with the end that you plug into your computer, your phone, your digital clock, your rechargeable bicycle headlight, or your electric toothbrush, you can't get the kind of shock I just described from these small wired boxes. These devices use a very small fraction of the power or rate of energy delivery compared with how much your wall socket could give you. (You can even repeat my childhood experiment with them, as I have accidentally done many times since.)

Your small charging units or power packs convert the power line's voltage to a much, much lower voltage, and then your appliance only draws or lets flow as much energy as it needs. It's elegant, and it's done with the extraordinary relationship between voltage and current, between volts and amps. Intuitively you know the bigger the hose or pipe, the more water it can carry. And so, the bigger the wire, the more current it can carry. But get this: The bigger the piece of iron, cobalt, nickel, neodymium, or samarium, the more magnetism it can carry. Iron is far and away the most common of these elements. I mean far and away. In these materials, magnetism flows like water through a pipe or electricity through a wire. We use the expression "magnetic flux" flowing through a "core" to describe it. It's a field of magnetism in another type of nature's pipes.

Each of these materials can carry magnetism. The bigger the cross section of stuff, the more magnetic flux can flow. It's another feature of physics, or nature, that makes sense, but only when you really take that moment to absorb the idea. In a wall power pack, we wrap the insulated wire coming out of a wall socket or receptacle around a piece of iron (or transformer steel, which is iron with just a schmink of carbon and a dollop of silicon). That coil of wire induces, or brings into existence, a magnetic field in the iron. We let that flow to another part of the iron piece and wrap this second area with another coil of wire. It works so elegantly, it's hard to believe at first.

If we wrap the wire from the wall around the iron 20 times, and we wrap the wire leading to your mobile phone around the iron once, the phone will receive 1/20 of the voltage that's coming out of the wall. If the wall is at 120 volts, the phone will get 6 volts. By controlling the number of wraps or turns of wire around our core, we can control or direct the electricity to be produced at a desired voltage and current.

Remember, though, that in nature there is no free lunch. You can't get out any more power, any more voltage-times-amperage, than you put in. And there is always, always some energy lost to heat. You've no doubt felt the warmth of a power pack after it's been in the wall for a while. This loss is attributable to the continual change of the electric field in the alternating current from the power company. The magnetism in a transformer is working or exercising the molecules in the core material. There's a wonderful term in materials science and physics for this: magnetic hysteresis. Molecules don't return to a given state on their own. They have to be pushed in both directions. Hysteresis just means that the magnetism in the metal converts some of the magnetic energy to heat.

The steels used in transformers are highly specialized. Researchers—metallurgists in this case—have found that by adding that little bit of silicon and rolling the steel out in very thin sheets and foils, they can greatly increase the efficiencies of transformers. In the way the steel is rolled and handled, metalworkers can decrease the losses in transformers by 30 percent. I mean, that's huge. It would be like having thirteen dollars in your wallet instead of ten dollars, just because you were careful not to crease or fold the corners of your bills. Transformers are everywhere; the less we lose in them, the better off we'll all be. It's another area where research might lead to breakthroughs.

Along with losses to heat, we have big losses of electricity just because we leave so many devices plugged in all the time. Until recently, some devices, like my old television, required 10 watts even when they were nominally turned off. Today there is a movement to require any device to use less than half a watt in standby mode. It's the power that keeps your television tuner, or microwave oven, or coffeemaker's clock running. With billions of appliances, we can easily waste hundreds of billions of watt-hours. This is sometimes called the "no-load" power loss or, what I prefer, the "vampire" loss. Each plugged-in plug has two prongs that are like the fangs of a vampire, sucking our electrical life's blood out of us all. *Bah, ha, ha, ha, ha . . .* It seems like an important problem to address, and we have, to some extent. Having vampires suck half a watt instead of 10 or 15 watts represents great progress. Nevertheless, how about if it were a tenth of that half? We could have billions of kilowatt-hours available to us for free, without any other decisions about nuclear plants or wind-turbine sites.

This ability to step-up or step-down the voltage is a key to our whole industrialized society. It's how we make so many devices that use so many different amounts of electricity go. We generate power at some pretty high voltage, say 500,000 volts in the turbines of the larger hydroelectric dams. We send it along power lines. Then we step it down 880 volts to send it to neighborhoods. Then we step it down at least once more before it gets to your house or apartment. Without the ability to transform voltages and currents, you wouldn't recognize our world.

So far, I've described a few inherent inefficiencies in the way we move power around. As I write, an overwhelming fraction of our power is produced by spinning machines. The spinning generators in coal-fired plants, nuclear plants, and hydroelectric dams inherently produce AC, alternating current that engineers describe by imagining waves. These are connected to synchronous spinning machines in our subway trains, escalator motors, and electric can openers. The waves also have capacitance with the surface of Earth itself. That's part of the reason the lines are strung up so high in the first place. But what if we didn't have those losses? What if there were another way? Or, better yet, a few better ways?

Over the last 150 years, we have set up our whole electrical grid world to run on AC. This goes way back to the beginning of electrical transmission. You may have heard the mythic stories of the competition between Thomas Edison and Nikola Tesla, "the war of currents." Edison advocated for a DC system. Tesla advocated for an AC system. Transmitting point to point, we would expect DC to be a bit more efficient. But practically, each end of the grid would have to be at the same voltage. The world embraced AC because of the enormous advantage it has, the ability to step voltages up or down at either end of the system with transformers. At Grand Coulee Dam, for example, we produce electric power at 500,000 volts. We transmit that over enormous distances, and step it down a few times, before it gets to your house or business. You've probably noticed the large can-shaped transformers hanging on power line poles. They are stepping the voltage down that one last time before it gets to your house or apartment.

Since we produce and transmit energy in three sectors of the circle around our generators, we can connect to a power line in one pattern to produce 110 volts for your house, or in another arrangement to get 220 volts. If you have a heavy-duty clothes dryer, it is almost certainly hooked up with 220, while the rest of your residence is running on 110. It is electrical engineering; it's elegant.

In the end, our society embraced Tesla's AC ideas, but the company that ended up dominating the electricity production market is named for Edison. No matter which side of the science history debate you find yourself on, Edison was a more successful businessman than Tesla. Edison died wealthy. Tesla died poor. Both were brilliant. But I wonder: How brilliant a person are you that you could invent this astonishing electrical transmission technology, but you could or did not notice that you couldn't eat? I find this bit of our history troubling either way I look at it. Was Edison an unfair businessman, or just a good one? Was Tesla so busy being brilliant that he was short on common sense? I can't tell. The historical accounts conflict. In any case, today we use both AC and DC, each where each is appropriate.

Back to our power story: Stepping up the voltage is like stepping up the pressure in a hose or pipe. If you wanted to get higher pressure, you squeeze the hose nozzle. If you want more flow, you open the nozzle. But either way, you have the same amount of energy and power available. You can't just say, I want higher pressure and higher flow rate, without adding power. The same is true in power transmission. The frequency or spin rate of the electromagnetic field is still determined by the speed of the spinning generators or dynamos. But transformers enable us to configure that power to have high or low voltage and low or high amperage. Compare this with Edison's original idea of DC electrical distribution. The farther you get from the power station, generally the lower the voltage a customer is going to get. With DC, there's no stepping up or down. Edison's customers got what they got, and that was that.

With this said about alternating current, when we go to transmit it over huge distances, and at high voltages, we still have a big problem: We lose power to capacitance and heat. So in the case of long-distance transmission, it's actually more efficient to send power as direct current. We do not have to deal with the enormously strong electromagnetic fields getting their energy sucked up by Earth's surface. We just have to insulate the DC power lines enough so that effectively no or very little energy gets lost to capacitance. To understand insulation in this case, imagine a wave stretched and stretched so that it becomes a straight line, infinitely long; that's DC power. Imagine capacitor plates infinitely far apart. No charge or energy can inadvertently build up on the plates; that's a way to imagine DC running in a big power line. There are a few dozen enormous DC power lines around the world, and indeed they are more efficient than they would have been if they were AC power lines. But still at some point, somebody has to convert that huge direct current to a huge alternating current. It's not trivial.

DC power does not solve all our problems all at once. To put it in an ironic way, we have to manage the management of power. At each end of a DC transmission line, we have to convert AC and DC. At the hydroelectric dam, we have to convert the huge voltages and wattages being produced by spinning turbines into direct current. For this we use very large electrical devices that reverse the direction of the current flow in an instant. Originally, engineers created electronic tubes filled with a tiny bit of gas like neon, xenon, or mercury vapor, which acts like an insulator in the tube until a certain voltage is reached. Then zap, the gas ionizes and lets the electricity through. They were called "thyristors" (such a word). Nowadays, we use devices called silicon-controlled rectifiers (SCRs) to do the same job. They are absolutely essential to an efficient power grid that uses a DC power line.

No matter how we transmit power, AC or DC, we are still working with metal wires strung up high above the ground. In any conductor, good as it may be, there is some electrical resistance. When we send electricity through a wire, we lose some of its energy, and that lost energy becomes heat, nothing but heat. When a material like metal gets warm, it softens. Put one candle in the freezer and another one on the kitchen counter. After a few minutes give each a squeeze. The warmer one is softer. So it is with power lines. When they're warm, they sag under their own weight. When they sag, their cross section gets thinner. It's just like stretching putty or pizza dough. Roll the dough into a rope. Then give it a stretch. It will thin and sag.

Most power lines nowadays are made of aluminum with either a steel wire or a fiberglass strand running through the middle for strength. Old lines like those at Hoover Dam are copper, which is indeed a better conductor than aluminum, though much more expensive. People steal it from construction sites because it's become so valuable. The other thing about copper is that it's heavy. If you're stringing power lines hundreds of meters over hills and dales, the weight of the wire becomes quite significant. The towers that support them have to be bigger, and the equipment to install them becomes that much more massive. At every turn, there may be opportunities to do things better. Even a little better will add up to great benefit, because we use so much power all day, all night, all the time.

Then there's another feature of moving electromagnetic fields that has to be taken into account. The fields and their energy pass through the outermost portion of the conductor. The energy moves along, almost entirely on the outside of the conductor, as though traveling through the "skin" of the wire. This came to be called the "skin effect." AC power lines are shaped to compensate for the skin effect, but any conductor has a surface, a skin, so it's not an easy thing to design around. At Hoover Dam in 1935, the state of the art was to manufacture large diameter copper cables that were hollow. It was a feat of manufacturing, and it was expensive. Nowadays, transmission lines handle the skin effect more cost effectively by having a pattern of wires suspended in groups. If there are three sectors of the power wave going down the line, each of the three is sent through a bundle of wires. They run right along together in parallel fashion, grouped like unbraided hair. It's much cheaper than a hollow conductor, just not quite as efficient.

There is triple trouble built into the way we do things, though. Let's say it's a hot day. Electricity consumers like factories, offices, schools, and houses crank up the air-conditioning and demand more power from the electrical power lines. As more energy goes through the lines, they get warmer than they were before the big power draw. As they warm, they sag. As they sag, they thin. So the electrical current has to pass through a thinner wire. Furthermore, the electricity has to pass through a skin whose circumference is smaller than it was when the wire was cool. So the wires get even warmer and even thinner, and so on. Designing for big loads on hot days is an essential and troublesome feature of modern major alternating-current power transmission lines.

As I mentioned in the last chapter, the U.S. Energy Information Administration estimates that we lose at least 6 percent of our electrical energy, enough to run entire states and provinces. It turns to heat that radiates into outer space. What if we could develop power lines that didn't have that feature, that just don't behave this way? What if we could take advantage of some other law of physics? What if instead of losing about 246 billion kilowatt-hours to heat up the air—and that's just in the United States, over just one year—we found a new way to address this problem and change the world?

Check it out. There really is a whole alternative field of physics that might let us do just that. If you've ever had a magnetic resonance image (MRI) taken, or been around the machine, you've heard that *bang, bang, banging*. It's a pump that keeps liquid helium cold. When wires like copper are that cold they conduct electricity with essentially zero resistance: Run an electrical current in them, and it just keeps going and going, seemingly in defiance of all the rules I just explained in such loving detail. The phenomenon is called "superconductivity." It's how we can maintain electrical currents high enough to induce magnetic fields strong enough to keep particles going at nearly the speed of light in the particle accelerators at the European Center for Nuclear Research (CERN) in Switzerland.

Recently, a 1-kilometer-long superconducting power line was set up between two power stations in Essen, Germany. It's a so-called high-temperature superconducting cable. High temperature is a relative term, however. The power line still needs to be chilled down to about $-140°C$. That's almost as cold as liquid nitrogen, but it is much, much warmer (if that word suits) than liquid helium. This operating temperature simplifies things. It works great. But we have to keep the whole thing cold, and that takes some electricity. Without the cooling, it will instantly turn to a hot wire, like you see in a toaster, only big enough to vaporize the wires and the coolant in a flash, and cause a good-sized explosion in this pretty good–sized city of half a million people.

Another way to improve the system is to tap into the amazing properties of the element carbon. In 2005, I met Rick Smalley. Unlike many readers of this book, Smalley won a Nobel Prize. By his account, the astronomy department at Rice University had detected a molecule in interstellar space that they couldn't understand. Using spectrographic analysis of starlight, they concluded that carbon was present, and it resembled the light signature of carbon monoxide. The perplexing part was that this molecule wasn't quite like carbon monoxide. Smalley told the story of how he woke up at 3 a.m. and realized what they must be seeing. It was a carbon molecule all right, but it was not like anything anyone had found before. Instead of carbon and oxygen, it was and is all carbon. And its atoms are arranged in a sphere, just like the pattern of leather swatches that make up a traditional soccer ball—white hexagons sewn with black pentagons.

Smalley called these molecules buckminsterfullerenes after Buckminster Fuller, the famous architect known for promoting geodesic domes and lightweight structures with elegant geometric connections. Smalley's dream was to create tubes of carbon atoms the same diameter as the bucky balls (as they're often colloquially called). By pulling hemispheres of bucky balls apart in a soup of properly prepared carbon atoms, his lab was able to grow tubes of carbon just a few nanometers across—"nanotubes." Smalley and his colleagues made careful measurements. These molecules become 10,000 times stronger than steel, and they weigh only a sixth of a similar piece of steel. They are amazing.

Wait, there's more. Smalley realized that if we could make power lines out of this material, with tubes nanometers in diameter but meters, or miles, or kilometers, or thousands of kilometers long, electrons would behave in a dreamy fashion. In a nanotube of carbon, it's as though electrons fall asleep at one end of the tube, have a dream, and wake up at the other end with hardly any electrical resistance on the journey in between. It's a consequence of quantum mechanics, the physics of subatomic particles and their strange interactions.

Right now, we can grow nanotubes only about 50 nanometers long. Smalley's dream is yet to be realized. But what if we pour our intellect and treasure into it? What if we solve the engineering problems associated with these nanotubes for power lines? It would change the world. My friends, there is no shortage of carbon around here. We could quickly dismantle existing lines, recover that valuable aluminum and steel, and greatly reduce the cost of transmitting power. It would change the world. Like Smalley's electrons, we can dream. And like Smalley said to me ten years ago, the key to the future is not to make do with less; it's to do more with less. Not wasting electricity heating up Earth's soil with capacitive transmission could be a huge step made with tiny tubes of ubiquitous carbon.

What if this technology could be applied everywhere? What if every wire in your house had virtually no resistive loss? What if every wire that runs every subway or commuter train did not need to be continually boosted just to get the power from one end of the track to the other? It would amount to way more than 6 percent. I can imagine it might be two, three, even ten times that. As the old saying goes, there are two ways to become rich: Get more money, or need less. If we needed less to do all that we do now and more with our energy, we would enrich people the world over.

# 16

# DUDE, WHERE'S MY BATTERY PACK?

Whenever I used to give someone a ride in the General Motors EV1 electric car, they would get what people in the electric vehicle community call "the EV smile." In the glove box I kept a logbook with the name of everyone to whom I gave a ride, and I really did sketch a smiley face next to his or her name in the book. The smiling response was invariable. After you drive an electric vehicle, you won't want to drive any other kind of car. Of course, as you probably know, there's a limitation that stops most drivers. Right now, EVs—electronic vehicles, that is—don't go far enough on a charge. But that is changing. My EV1 barely went about 80 kilometers (50 miles) on a single charge. I recently drove a Tesla Model S that can go over 500 kilometers (300 miles). We're in the midst of a big good trend.

General Motors seems to have built the EV1 to satisfy a law in California. After GM successfully lobbied, and California loosened its regulations, the company's executives chose to abandon their EV program in 1999. They had the much-loved vehicles literally crushed. That decision motivated the production of the movie *Who Killed the*

*Electric Car?* Perhaps it was the same thoughtless managers who mysteriously named the prototype version of the EV1 the "Impact." When I heard that name, I could not help but hearken to the words of my parents, "Common sense is not that common." Why not call it the "Crash"? I can hear the marketing person explain that Impact is a good name ". . . because, y'know, it will impact the market and crash through with disruptive concepting . . ." But I digress.

At any rate, the EV1 was a sexy car, and not just because it looked geeky cool. It also offered a wonderfully different relationship between the automobile and the energy that goes into it. If you drive a conventional gasoline car, your only option is to fill it up with gasoline. If you get a diesel car, or even one that runs on compressed natural gas, the situation is the same: You are locked in to using carbon-spewing fossil fuel, and the only thing you can control is how much or how carefully you drive it. In an electric car, though, all that matters is the electrons . . . and every electron is exactly the same (as far as we can tell). If you recharge an EV with electricity that was generated from coal, you are basically plugging into a coal-fired plant. But if you recharge using electricity generated from wind and solar, right away your EV becomes a true carbon-free vehicle. Best of all, if your local utility keeps getting cleaner and cleaner, your car keeps getting cleaner and cleaner, too. You don't have to change a thing; you're just minding your electrons.

So you can see, the batteries were the best thing about the EV1. Unfortunately, they were also the worst things about it. I drove the first generation battery pack for several months. That 80-kilometer range was a pretty hard limit, and that was if you were driving conservatively. The instrument panel was elegant and the motor nearly silent. Driving the EV1 felt like you were zooming even with gentle accelerations. But in the Los Angeles area, a 40-kilometer trip one way is not uncommon. I routinely planned round-trips from which I could not return unless I could find a charging station along the way. The energy storage was a serious limitation. So was the cost of the batteries, which reportedly made the EV1 a money loser for General Motors. Being charitable, I guess that combination spooked GM management, so they gave up.

Other companies are now stepping in. The idea behind the electric car is just too good, and the technology is rapidly improving. I recently leased BMW's Mini Cooper Electric and it was, as the kids say, "all that." I ended up driving it for a year. I was a test driver on the Chevy Volt. It is the best GM can manage so far, having been compelled by those dingbat managers to abandon the huge head start they had with the EV1. I had a Nissan Leaf for three years; it is a lovely little vehicle. So is the peppy Volkswagen E-Golf. I drove it and loved it. The Ford equivalent, the Focus Electric, is a winner as well. The BMW i3 is another hot rod. It has an optional, very small gas engine to supplement the batteries. I loved driving it. The Tesla is the real deal, but it's also really expensive—at least for now.

To rev up—wait, I mean to energize any of these cars—you push a single button. I commuted in the Leaf, the Golf, and the i3 to and from the Planetary Society, a 50-kilometer round-trip. That consumed about 10.5 kilowatt-hours (kWh) of energy. One big adjustment with electric vehicles is thinking about their "fuel" the way you think about your electric bill rather than the way you pump gas. On that commute, I never had worry or anxiety about the range provided by the onboard energy-storage system—the batteries. But if I made the trip in the Leaf from my place to LAX (Los Angeles Airport) and then on to the Planetary Society, I always got a little anxious. I would seek a place to charge. One night, I pushed it all the way down to less than 2 kilometers of range left. With all the batteries on board, these electric cars are quite heavy. Nevertheless, I had it in my head that I could push my car the last few soccer field–long city blocks. I made it home that night under battery power, but I suffered badly from range anxiety. Some additional disclosure: I recently pulled the same stunt in the E-Golf; it's fun in its way, but it's absolutely not the best thing for your relationship with your passengers. They want to get home.

To virtually eliminate any range anxiety, the BMW i3 has a 50-mile range on gasoline. It's just for emergencies, when you need to get yourself to a charging station. I tested it; you can hardly feel the switch over to gas. In the fast-moving Tesla, that same 50-kilometer commute would have consumed 14.6 kWh of electricity, an increase of about 25 percent, but the Tesla more than makes up for it by packing in a much bigger battery pack. More batteries mean more range, but they also mean more weight and cost. The weight increase reduces efficiency a little, but I think most people will happily trade a little energy performance for added range performance. The cost is a tougher trade-off. The Tesla Model S is a remarkably luxurious car, though. It's pretty stiffly sprung, like a sports car, and it has a selectable operating mode that is literally called "Ludicrous." The car outperforms any comparable internal-combustion vehicle, doing 0–60 in about 3.2 seconds. You burn through kilowatt-hours, but wow what a ride!

Because their drivetrains are so efficient, and there are no explosions to muffle, electric cars are all so much nicer to drive than any internal-combustion vehicle. When I talked on the phone in the Leaf or the E-Golf, or i3 hands-free, people thought I was holding a phone to my ear because the cars are so quiet. After an electric car, other vehicles just seem like old-tech, gas-burning, pollution-making throwback machines. Which, frankly, they are. But if we each had an electric car tomorrow, there would be a problem. Our power grid could not handle everyone drawing an additional thousand kilowatt-hours every week. That sounds like a serious challenge but, as with so many issues surrounding energy and climate, the challenge is also an opportunity. We could, if we had things figured out, make electric vehicles

a new component of the grid and use them to store energy for everyone in everyone's car.

You see, the range-limitation issue is connected (electrical pun intended) with a much bigger problem of energy storage for solar and
wind power—really, for any energy source that is not constant. If we
can address the energy-storage problem in cars, using technology that
exists or nearly exists, I believe we can address the problem on a huge
scale for our energy supply writ as large as can be.

Engineers and city planners know just about exactly how many
cars are going to be parked in what part of a city or suburb at just about
any time of day. (It reminds me of the eyebrow-raising statistics concerning the number of toilets flushed at halftime of the Super Bowl.)
With this knowledge of where vehicles are parked, a carefully designed automated electrical grid could move energy for everyone
around from parking lot to garage to parking lot to shopping mall,
etc. We could use the idle batteries of everyone's car; those people
who knew that they had spare range available in their vehicle's battery
pack could select a "store credit mode" and have the car automatically
notify the software of the smart electrical grid to get credit on that
owner's electric bill.

It may seem like a complicated plan, but it's really no more complex than a system that enables you and me to travel from one mobile
phone cell to another with the calls being handed seamlessly from one
tower or antenna to the next. If we were motivated, we could do this.
We could fund battery development. We could fund research into
lightweight, high-strength materials to make cars lighter. We could
provide tax incentives everywhere and make the transition from gas to
electric vehicles on a large scale.

The goal here would be to power everyone's electric vehicle with
renewably produced electricity, from wind and solar energy, say. But
even if the power is generated exactly the way it is now, with nearly

half of the power in the U.S. coming from coal, electric vehicles have advantages for the environment. First, they are not constrained by that Second law of Thermodynamics' heat engine. The energy-generating power plant is, but it can be optimized much more readily than can a car engine that is required to continually change speed and operating temperature. Second, the pollution at the power plant can be controlled or filtered. The exhaust gases are not dragged all over the map by commuters and soccer families.

To get started on all this, we could let the price of gasoline rise to what the market would actually bear. Today, gas in the United States is subsidized to hold its price at about half of what it is almost everywhere else. Although the price of gas or petrol has been a lot higher in the recent past, right now it's about $0.75 a liter ($3.00 a gallon) in the U.S. In other countries, such as Britain and Japan, people routinely pay about $1.20 a liter or more ($5 a gallon). But they still drive. So many people in the U.K. try to drive into London every day that there is a £11.50 ($18.00) "Congestion Charge" just to cross the city limit in a car. And people pay it—every day! Some people want to be in cars that badly. More on that later.

I'm sure there are some readers who will choke on the word *subsidy*. The gasoline subsidy in the U.S. is not a visible thing that shows up in the budget documents, but it's there. Keeping a standing Army and Navy at the ready to defend oil fields on the other side of the world is an extraordinary subsidy for gas-powered vehicles. Cheap leases on federal land for drilling and mining are subsidies. Tax breaks for the fossil-fuel industry are subsidies. If we enhanced the grid, subsidized electric vehicles, and let gasoline cost what people are really willing to pay (and what we really pay to get and protect it), we could bring a lot of our military home. And change the world.

As I write, the biggest perceived enemy of the U.S. is called the Islamic State of Iraq and Syria (ISIS), or alternatively the Islamic State of the Levant (ISIL). They are bent on taking over territory by means of terror. If the Western countries did not need the oil that these terrorists control, the money that ultimately funds their operations would dry up. They could no longer operate as a terrorist state, certainly not at the level they do right now. Protecting large populations from the thuggish ISIS would still prove complicated, but not nearly as complicated as it is now. Subduing ISIS will require international cooperation, both military and diplomatic. Freedom from oil would make that task easier. Fossil-fuel energy is the cause of the conflict and the enabler of the terror. The whole web of American involvement in Iraq, Saudi Arabia, Syria, and other regions of the Middle East is tethered to petroleum, either directly or indirectly, through our political ties back to the countries that have the oil under their soil.

For me, this is all the more reason to work toward a sustainable, post-fossil-fuel-energy economy. As the saying goes these days, "Haters gonna' hate," but if we don't need what they have, they will have a lot less ammunition for exporting violence, and we will be a lot less exposed to it.

# 17

# QUEST FOR STORAGE

In the U.S., where a great many young people grew up watching the *Bill Nye the Science Guy* show, passersby often ask to take a picture with me. I'm generally happy to do it, but often our technology gets in the way. If I had a nickel for every time someone has asked me for a picture and discovered his or her battery was dead, well, I'd have a few dollars. Batteries have become a leading constraint for a lot of our modern tech. These days, the main limitation of how long a U.S. soldier can last in the field is no longer the amount of water he can carry in his canteen but the amount of electricity stored in his batteries. When he's out of juice, a soldier is stuck, isolated from his commander, effectively deaf and blind to what is happening in combat. In my years of doing TV I've walked past barrels of discarded batteries: Shows use batteries to power the actors' wireless microphones, and we don't want the mic to fail mid-performance. So the problem of weak batteries isn't new, but with the increasing importance of renewable energy and electric vehicles the problem has become a lot more pressing.

The Tesla Corporation—the folks who make that wild Tesla

Model S—is getting ready to introduce battery packs for the home. They're supposed to be available in 2016 for a cost of a few thousand dollars each. These are lithium-ion batteries, similar in technology to the ones in today's electric vehicles, bundled into a unit about 1.3 meters high and a meter wide. They can each store about 10 kilowatt-hours. Once charged up from a house's solar panels, two of these units could run most houses all day long in the spring and fall. (By the way, a standard Tesla car has a 60 kilowatt-hour battery pack. In other words, it takes three or four times a household's daily energy requirement to drive a state-of-the art electric car on the highway for a few hours.) That's a good start, but still far from what we'll need to remake our energy supply around clean but inconsistent solar panels and wind turbines.

Not to put too fine a point on it, but if you can invent a better battery or electrical storage system, you'll get rich, crazy rich. Because of the astounding potential (a pun), a great many researchers are doing quite a bit of research to come up with a more compact way to store electricity using chemistry. Alessandro Volta developed chemical piles. These were stacks or piles of dissimilar metals kept separate from each other by layers of brine-soaked cloth. You could make frog leg muscles jump or twitch just by touching the muscles with the two leads or wires. The key to a battery pile, Volta realized, was to have two different metals. One hundred and fifty years after Volta's early discovery, chemical engineers the world around are still working with dissimilar metals and other materials to improve our batteries. We want batteries to hold more and more energy yet weigh less. Is there something better? Maybe not. Maybe we need to do more with the chemistry we have.

When I look at one of Volta's original piles, it looks a little sloppy. He was using the plates or wafers of metal that he could get. But seeing them still throws me off a little. The not-quite-so-flat surfaces must have severely limited the amount of current and voltage that thing could produce. Modern batteries are the same way. It's not just the chemical reactions as such. A battery's performance depends on its geometry, how the chemicals are arranged mechanically. There are still layers. There are still solvents. There are still metals, just like in Volta's day. To get them to perform well, a battery's electrodes and chemicals have to be arranged and manufactured carefully, and to close-fitting tolerances.

A handy idea when it comes to batteries is to compare a ratio of how much energy, how many Joules, or watt-hours a battery can hold to how big it is, how much volume it takes up, or how much it weighs. Strictly speaking, a battery's "energy density" would have to do exclusively with its energy and volume, and another phrase "specific energy" would have to do with a battery's mass or weight. But by long tradition, "energy density" is often used for either. You have to see which units of measure are involved. In either case, in general, the higher the density, the better. Note well, the bigger a battery is, the lower its energy density. And the heavier a battery is, the lower its specific energy (or energy density by weight). Either way, more energy is better.

In the same way that the food you eat has chemical energy, the chemicals in batteries have chemical energy. Electricity is the flow of electrons, and our goal with a battery is for the electrons to flow from one electrode to the other. In general, the electrons flow from the negatively charged side to the positively charged side. In a battery, we say the electrons flow from anode to cathode. But oh my friends, what we call the "electric current" flows the other way—from cathode to anode. Humankind did not create this confusion on purpose. It pretty much dates back to Ben Franklin himself, who did his best to determine which way the electricity was flowing using static electricity. Without access to balloons to rub on his hair (oh, the humanity), he was constrained to use rubber rods, glass rods, fur, and silk.

Put simply, Franklin guessed wrong. He thought the flow of particles of electricity, if there were such things, went from cathode to anode. It turns out to be the other way around, and here's the surprising fact of nature. The description works fine either way. Analyzing circuits as having a positive charge flowing from positive-to-negative turns out to give exactly the same result as analysis that presumes negative charges are flowing from negative-to-positive. After fifty or more years of thinking about this, it still surprises me a little. Knowing which subatomic particle is flowing in which direction, however, is quite important when we go to understand batteries because in batteries it's all about the ions (from the Greek word for "go").

If we say to ourselves: we're going to have a flow of electricity, in this case, carried by electrons, and those electrons are going to be driven by the energy in chemicals, well, it implies that to complete an electrical circuit, we're going to move some chemicals around inside this battery. So herein lies the rub—by that I mean the friction. Getting chemicals to drive electrons means that some chemicals are going to have to move past each other in the container that is the body of the battery. The chemicals of interest are suspended in a liquid or gel, and after Michael Faraday we call the mixture the "electrolyte."

Volta's piles used copper and zinc. If you're scoring along with us, and I hope you are, those metals are right next to each other on the periodic table of elements. Copper has 29 protons. Zinc has 30. In the right solvent or electrolyte, like salty water or sulfuric acid, electrons move from the zinc to the copper, while atoms of copper move through the liquid and stick to the zinc. This is how we coat or plate metals with electricity; it's just called electroplating. I mention this because Volta's pile was the first chemical battery.

You may remember carbon-zinc batteries. They are heavy and distinctive because the bottom of the battery is the dull gray color of zinc. These batteries rely on the chemistry between zinc metal, ammonium chloride, and manganese oxide. A carbon rod is embedded in the middle and is a good conductor that doesn't get ruined by the reaction inside. A very similar reaction takes place in the popular alkaline batteries. It's still zinc and manganese oxide. But instead of the ammonium chloride, these cells have potassium hydroxide. It may sound a bit chemically complicated, but just glance at the periodic table. All of these reactions involve two chemicals: One is a metal and the other is something to make oxygen bond with them.

In general, the electricity-producing reaction forms a metal oxide, and produces some unused waste product or compound along the way, which has to be contained or immobilized in the electrolyte's paste or goo, along with the chemical that's driving the reaction in the first place. And while these chemicals are moving one way through the goo that is the battery's electrolyte, electrons are being driven from the anode to the cathode. The chemicals that are in motion either have more electrons or fewer electrons than normal traveling with them. These are the molecules we call ions. If it's a conventional flashlight-style battery, electrons are going from the wide bottom to the narrow button-shaped top. And that in turn means what we define as electrical current is going the other way, from top button to wide bottom, the same way the ions are moving and oxidizing the metal can or outer metal container of that style of battery.

The battery business is competitive. We humans in the developed world go through billions (with a b) of batteries every year. It's a huge market. If engineers can gain even a slight advantage in manufacturing batteries, they go for it, because we all want to store electricity in manageable compact batteries. There are all kinds of them: nickel-cadmium (NiCad), nickel-metal-hydride, silver-zinc, mercury-zinc (mercuric-oxide), and zinc-air, to name some popular ones. Batteries have an energy density. At the low end, lead-acid densities down around 80 watt-hours per kilogram. NiCad is about twice that. Then densities go all the way up to zinc-air at over 400 watt-hours per kilo. The very popular lithium-ion batteries weigh in (ha!) at up to 250 watt-hours per kilo.

That last one is the type of battery that is generally the state of the art right now: lithium-ion. In general, these batteries exploit a reaction of lithium, cobalt, and oxygen. Because lithium has only three protons, it's lightweight. It makes for a small atom and a small ion. The crazy thing for me, though, is that these work so well, not just because of the shape of the can, or the container, or the cell, but because of the shapes of the molecules involved. They form octahedra (like two pyramids smooching base-to-base). There's oxygen on either side of the pyramids. It bonds and un-bonds easily in the midst of the octahedra. By the way, as I write, there are researchers claiming that through clever geometry and chemistry, they can increase the energy density of lithium-ion batteries by a factor of seven. The improved density would be up around 1,700 watt-hours per kilo. That could make electric vehicles wonderfully desirable. If this research proves out, it could change the world.

Lithium-ion batteries are great. But if these batteries get hot, their oxygen can get loose. A few of these batteries have been known to catch fire. Yikes. You might say, why don't they put something in there to absorb that oxygen if it gets too loose? Well, anything mixed in with the electrolyte is almost certainly going to lower the effectiveness or current-producing ability of the whole assembly. It's a mechanical-electrical-chemical engineering trade-off. After years of engineering, fires seldom start. But it's a real issue that has to be addressed with every design. Battery fires slowed the introduction of Boeing's 787 airplane. In fact, the Federal Aviation Administration (FAA) temporarily grounded the whole fleet of these more than $200 million planes in 2013, because of trouble with a few lunch box–sized batteries. (All of the 787s are back in service now.)

When any of these batteries are kaput, said and done, depleted, used up, we have converted a metal to a metal oxide. I had a professor who said it's like we're burning the zinc, oxidizing it. This idea is being expanded to make more energy-dense, more efficient, more easily recharged batteries. The Tesla Corporation plans to recycle well-worn car batteries and convert them for use in consumer's homes. It's a good step with existing technology. But one of the emerging specific chemical reactions or technologies could change the world.

I don't know you, but I'd bet that you've been in a conventional automobile. Or you've at least seen them. Their batteries are what we call "lead-acid." If you've ever carried one, you know they're quite heavy. But they work great, and they have the wonderful feature of being easily recharged. You can let the current flow from positive to negative, with lead dioxide ions flowing from the positive to the negative side, and start the car. Or you can have the car's alternator drive electrons and lead dioxide ions back the other way, and recharge the battery. You're probably all too familiar with the hassle of recharging a discharged (dead) car battery—usually in the rain, and always when you are (or were) in a hurry to get somewhere.

Traditionally in these very successful and ubiquitous batteries, there is a series of cells. If you've ever even glanced at an old one of these car batteries, you probably noticed the cap over each cell. As the battery delivers its charge and gets recharged many times, some water evaporated; some sulfuric acid got chemically bound to the lead. On some of these old batteries you could get quite a bit more life out of it by replenishing the water, because even if the chemical system is degraded a little, the key is still having enough liquid or gel electrolyte around that enables the ions to migrate. Nowadays these batteries are sealed, and the electrolyte is a stable gel. To reuse the material inside they have to be disassembled with special equipment. It can be done, and it's good business.

In traditional lead-acid batteries the cells are connected to one another in a series, and those connections almost always involve something like solid conductors, pieces of metal hooking one cell to the next. The arrangement works fine. But in the last few years engineers have come up with a glasslike ceramic coating that enables the negative plate of one cell to act like the positive plate of the next cell. There is no need for the jumper conductor. The whole thing becomes much more compact, way more compact, almost half as big as conventional batteries. This ceramic plate between the sheets of metal changes the shape of the battery, while the chemistry is pretty much the same. Since the ceramic is acting like a positive pole and a negative pole at the same time, these have come to be called "bipolar" batteries. Just imagine if every battery you ever came across were half as big as it is now. It could enable us all to do more with less in a great big, half-the-size, kind of way.

You probably noticed that each of these styles of batteries is just like us; the chemical reactions that make them go involve oxygen. Regular ole air may be a key to future batteries. Carefully designed batteries can get their oxygen not from another oxidizing chemical, but right out of the air. And why not? The planet's atmosphere is lousy with oxygen. Can't avoid it, really.

There is a great deal of work being done right now on aluminum-air batteries, zinc-air batteries, magnesium-air batteries, and lithium-air batteries. These are set up so that the electrode that accepts incoming electrons (the positive terminal or pole) is actually just air, of all things. The energy densities of these batteries is, theoretically at least, about as high as gasoline. Just imagine. Instead of a fuel tank, or instead of a very large, as-big-as-the-car-is-long, battery pack, an electric car would have a battery no larger than those two five-gallon cans you may have seen on the back of jeeps. We could run every car in the world electrically. Wow!

After running these batteries for a while, the aluminum, in this example, is used up, converted to aluminum oxide. It hasn't disappeared. It's just been chemically changed so that you can't get any energy out of its battery. Replace that aluminum electrode, and off you go again. It could become a big part of the future. Then, in my imagination, we would use electricity produced renewably from wind and solar sources to recover the aluminum in easy-to-envision aluminum-electrode recovery plants, and reuse the aluminum indefinitely.

We pretty much do this already with the lead in lead-acid batteries. You trade in your old battery when you buy a new one. Furthermore, almost everywhere, you are required to trade in your old battery. Society (your neighbors) does not permit you to toss this metal away. It's valuable, for one thing. It would become a heavy-metal environmental pollutant, for another. Recycling lead and aluminum is in everyone's best interest. We should all just come to expect that batteries must be manufactured so that their key components can be recovered and re-used. It's a new way of thinking more than it is a way of imposing new regulations.

Ever since I messed around with electroplating and batteries as a kid, I've wondered about the metal. I remember being fascinated (I guess I still am) with carbon. The graphite "lead" in a pencil conducts electricity. I know, because I've done this many times: get a wooden pencil and sharpen both ends. Connect batteries and a lightbulb to it; disassemble a flashlight, for example. Electricity will flow, after a fashion, right through the pencil lead. So now, sure enough, people are experimenting and developing batteries with electrodes that are just rods or layers of carbon. Old chemistry can become new chemistry. Old batteries had one carbon electrode that did not corrode. In these new designs there is an electrolyte between two carbon electrodes. Both the cathode and the anode are carbon, so neither one corrodes. We're not burning metal. Carbon can be a bit brittle, but these batteries, if they can be made durable enough, could last and last. It could revolutionize how we store electricity.

When it comes to storing energy, electricity in a battery has the potential (pun again) to be about as good as it gets. I have no trouble imagining a day in which each application, whether it's a car, a boat, a factory, a skyscraper, your house, or your favorite airplane uses a different type or style of battery. If we really can come up with all these different designs and make them reliable, we could do more with less by storing renewable electrical energy everywhere it's needed. We could distribute the energy when it's available, and draw it out of our millions of distributed batteries when we need it.

The humble battery could be the key to the future for humankind. Is that a big deal? I sure hope so!

With all this battery business covered, I should remind us that a chemical system is hardly the only way to store electricity. When it comes to energy, it all adds up. Joules are Joules. Newton-meters are Newton-meters. Foot-pounds are foot-pounds. At a place like Ross Dam on the Columbia River, we convert the energy of falling water to mechanical energy, and then convert that to electrical energy. The backed-up water behind the dam is like a giant battery. It's sunlight that causes water from the ocean, lakes, and ponds to evaporate. Water vapor molecules are less massive than other gases in the air, like nitrogen. So up they rise until the pressure around them lets them cool off, condense, and fall as rain or snow. The falling is gravity doing its work.

Suppose we took advantage of gravity to store energy in another way. As I write, people are proposing enormous gravity systems that could store 100 billion Joules worth of energy, and return a few megawatts of power. And they'd do it by suspending a big ole heavy weight and letting it fall down. The details are actually pretty straightforward.

Picture an old-style cuckoo clock; you might already know how it works. To wind such a clock, you pull on one, two, or three chains to raise one, two, or three weights. They are traditionally shaped like pinecones (the ones native to the Black Forest in Germany). Then a mechanism called an "escapement" lets the weights fall very slowly, driving a back-and-forth motion featuring specially shaped gears. The system lets the gravitational stored energy escape slowly. Suppose we did such a thing on an industrial scale. Such a mechanism would be huge, and it would require some pretty heavy-duty gears and shafts and some kind of yet-to-be-figured-out escapement, but we could probably pull it off. The key would be making each mechanical component especially robust and stiff, so that the mechanism could operate for years and years.

With this said, I believe some thoughtful engineers have found a somewhat better way to harness the energy of an enormous falling weight—water. Not conventional hydroelectric dam water, but water as a working fluid in a contraption so simple and elegant, yet so huge, that it just might work. We would construct a very large vertical shaft in the ground where we wanted to store energy. It would, I imagine, penetrate the water table. We do this all the time with oil and gas wells. We line the inside of the borehole with a casing that is watertight. It can be done. No matter how you feel about wells and drilling, millions of oil wells have been drilled that do not affect nearby water wells. Although there are notorious leaks once in a while in fracked natural gas wells, when it comes to drilling straight and vertical, and sealing things up tight, drilling crews have that down.

In this very large casing we would build a piston—one huge heavy piston made of local rock, mined right there out of the rock removed to drill the shaft. We'd finish it smooth with one or more layers of concrete. If we lift this enormous piston, we would be converting lifting energy to potential energy. We'd lift it with water driven to the bottom side of the piston with a huge water pump. Then when we want the energy back, we let the weight of the piston drive the water back up the shaft and through the pump running in reverse or through a separate turbine and generator combination rigged up to optimize the flow in that direction. It's an idea so simple, it just might work—for years and years. If the pump breaks, replace it. If the seals around the piston leak a bit, it just lowers the efficiency a little. In that event, just pump it nearly dry and replace the seals. It would not lead to a catastrophic failure, as it might with a plumbing leak in a nuclear plant.

The other feature of this hydraulic gravity energy storage setup is that it would be scalable. We could build these systems big and small; I mean big or gigantically huge. Imagine a piston 30 meters in diameter, 100 feet across. To keep it stable, it would be at least that high or thick. If it were made of solid rock and concrete, it would be one-and-a-half times as dense as water. Such a piston would weigh at least 200 million kilos, or 200 thousand tons. We're talking more than four battleships. If we were to produce electricity nearby, at a wind farm or extensive solar array, we could drive that piston up all day, and let it descend all night. A piston such as this could store over 50,000 watt-hours. If we had an array of them, a piston farm, we could store a couple megawatts, the equivalent of a regular power station.

An important feature of this idea is just that it would be cheap. We have the technology to dig huge holes for mining. And at these mines we already push or haul the tailings away. In the case of these piston wells and their casings, we have all the technology we need right now. We blast holes in the ground and bedrock. We cast concrete structures. We have water turbines. We have generators. We just need to decide if it's worth a try near a wind or solar farm. If it works as well as it seems like it would, we could mass-produce these systems and put them in the ground and bedrock wherever we determine we need them most.

Now suppose you wanted a system like this for your house. Skilled crews could drill down just 20 meters. You could have a 10-meter high, 1-meter-diameter piston. When pumped to the top by your solar array—or by the arrays of your neighbors, who might have fewer trees blocking their solar panels—you could store almost 2 million Joules. That's only half a kilowatt-hour. But now, suppose it were 2 meters across (the size of a circular patio table), and the shaft were now 40 meters (130 feet) deep. You'd have over 8 kilowatt-hours, which might not cover your house's full-time needs, but it would sure run your refrigerator and get you and your family through something like Hurricane Sandy.

Now suppose, further, that everyone in the neighborhood did something like this, and these piston and pump systems were mass-produced and cheap. We could substantially cut our energy storage needs, and we wouldn't be dependent on unpredictable advances in battery technology. We might not have to rewire our old power line distribution systems, leastways, not right away. In ten years, each house would be drawing half as much power as it does today. These hydraulic piston shafts could become common enough to keep our cities running while we work to bury bigger, more capable power lines and build the associated smart distribution grids. It would be part of the big idea of doing everything all at once.

It is a strange but irrefutable feature of energy storage: The more we store, the less we'll need to produce when the Sun isn't shining and the wind isn't blowing. Storage would even reduce the load on nuclear power plants. Each plant would not have to be made quite as big. Their environmental impact would be smaller. It would reduce the risk of a nuclear accident. All this by pumping slugs of rocks up and down in concrete tubes.

Having gravity-storage pistons near local power plants might be an outstanding way to reduce the peak loads on those plants, enabling us to use our existing electrical-grid infrastructure while upgrading it along the way. Solar panels on as many roofs as we can manage would reduce the peak loads as well, especially if they are coupled with battery or gravity electrical storage systems right there at each house or building. The key is thinking about the long term.

The solar panels on my house in Los Angeles have paid for themselves, and that took almost seven years. Still, if we think in decades instead of two- and four-year election cycles, we can accomplish huge things, one battery or piston at a time.

# 18

# BOTTLING SUNSHINE WITH MOONSHINE?

Once you start thinking about storage, it's hard to stop. At least, it's hard for me, and I'm hoping soon it will be hard for you as well. Because energy storage is everywhere, it's no exaggeration to say our lives are all about energy storage. When the wind blows, that is solar energy stored as motion, or kinetic energy. When the rain falls it carries solar energy that was stored as the heat of evaporation. And all of our food is just bottled sunshine—bottled either with photosynthesis by plants, or sequestered by the animals that ate those plants. It's storage all the way down.

While engineers are working on clever new ways to store solar energy in batteries and maybe giant pistons, farmers are already doing it on a huge scale. A field of corn is a field of solid sunshine. Ferment that corn and you get energy in an even more concentrated form. You may be familiar with country songs that describe "white lightning," and "moonshine." You may also recall a line from what is now the University of Tennessee fight song: "Corn won't grow up on Rocky Top, dirt's too rocky by far. That's why all the folks on Rocky Top get their corn from a jar." It's Rocky Top, Tennessee. The singer is referring to alcohol made from corn—ethyl alcohol, or ethanol. You can drink it and feel the burn. Uh . . . so I've heard. I mean I would never . . . but my point is, you can set fire to it and it will really, literally, burn. You can watch the stored solar energy coming right back out.

With all this good stuff that can be fit into a jar, why not run our world on ethanol? If it were possible, we could replace fossil fuels with the wave of a cornstalk. Burning ethanol releases carbon dioxide, but making it pulls carbon dioxide out of the atmosphere. If you can be efficient enough, the two would nearly balance out, producing little net carbon emission. For cars, ethanol could clean up what goes into the tank while we are ramping up production of electric vehicles. And for other oil-dependent forms of transportation—container ships and airplanes, for instance—switching to ethanol-based fuels seems a lot more plausible than running them on battery power. A moonshine-based economy. Sounds cool, right?

If only things were so simple. The conversion of sunlight to corn to ethanol turns out to be barely 2% efficient. Doing a wholesale swap from oil to ethanol would require growing enough distillable vegetable matter to fuel the world for a year during just one growing season. This, uh . . . turns out to be somewhat complicated. I mean, not possible. We are constrained by the shape of the planet and the chemistry of the plants. Despite considerable misconceptions and pitfalls, though, there may be some opportunity here.

For a few years, ethanol and other plant-based fuels (collectively known as biofuels) got a lot of enthusiastic press coverage. Then more scientists started to step up and talk about the drawbacks, or at least the big challenges, of ethanol. First of all, gasoline gives you more than twice the combustion bang for your buck—or kilo, or liter, or pound, or gallon. Gasoline or petrol carries about 45.8 megajoules per kilogram (19,700 BTU per pound), while ethanol carries less than half of that, about 19.9 megajoules per kilogram (8,560 BTU per pound). A Joule is about the amount of energy required to lift a stick of butter from a floor to a table. A British thermal unit is the energy needed to heat 1 pound of water 1°F. If you've ever used a camping cooking stove fueled with "white gas," you probably have an intuitive sense of the difference in heat energy between these two fuels. Compare how hot a chafing dish gets at a banquet food line with how hot gasoline burns in a camp stove. The term "white gas" generally means gasoline with none of the many additives used nowadays to help car engines run cleanly and efficiently. If you have not experienced both of these combustion processes, I can tell you gas burns hot. Alcohol only burns . . . hot-ish.

Science fiction writers generally ignore this factor-of-two difference in the specific energy (energy per unit of mass) of gasoline and alcohol, which has contributed to unreasonable expectations. In the movie *The Rocketeer*, and in the more recent *Tomorrowland*, writers and directors expected the audience to believe that alcohol has almost magic powers. These guys were flying all over the place with what looked like just a few ounces of low-energy alcohol. That much alcohol could not mow most front lawns. It's a long way from the fictionalized flights of Cliff Secord, the rocketeer, and young Frank Walker, the inventor and time traveler bent on saving the world, to what we actually need to save the world. But then, what's science fiction for if not to make us dream? Still, both of the films I mentioned had a pie-in-the-sky quality that took me out of the story. Ethanol just can't do those things. Its energy is just not that concentrated.

Then keep in mind that growing enough of anything—corn, sugarcane, or whatever—takes a lot of land. Then the corn has to be harvested and distilled. There's a lot of work involved, and a lot of energy has to go in to get the energy out. Nevertheless, if you don't have to go drilling and pumping to get fuel, if you could just grow fuel on farms, it seems like you could do great things.

Try this: Let's say about 1,000 watts of solar energy makes it to every square meter on the surface of Earth where we have a farm field. On every 100-meter by 100-meter hectare of land, we get 10 million watts every second the Sun shines at full intensity. Over a season, that hectare receives about 100 billion Joules of jolt from the Sun. (Remember: Watts measure the rate of energy flow, Joules measure the total amount of energy.) At the current rate of converting about 10 metric tons of corn into about 1,600 liters of ethanol, we're looking at a conversion efficiency of sunlight to chemical energy up around 10%. But that does not take into account all the many agricultural activities and energy inputs needed to grow all that corn. The seeds have to be grown in the first place. They have to be transported to the field. They have to be planted with a tractor, watered, fertilized, insecticided (*sic*), and harvested by another tractor.

With all that, the efficiency of ethanol production goes way down—down to the 2% number I cited earlier. The U.S. National Center for Environmental Economics (part of the EPA, the Environmental Protection Agency) has cautioned that ". . . biofuels can emit even more greenhouse gases than some fossil on an energy-equivalent basis. Biofuels also tend to require subsidies and other market interventions to compete economically with fossil fuels, which creates deadweight losses to the economy."

In Brazil, cars are propelled generally with a mixture of gasoline and alcohol produced from sugarcane grown on the country's farms. Sugarcane also empowers drinkers to set sail with a Captain Morgan and brewers to distill rum that can make for delicious mojitos. (That's what I've heard, anyway. . . .) Brazil has enormous tracts of arable land, and much of it is in the tropics, where the very sugary sugarcane can thrive. There is more sunlight to convert to more sugar-rich fuel. You get about twice as much alcohol out of a hectare or acre of land in Brazil, growing sugarcane, than you do in the U.S. growing corn.

Sugarcane thrives in the tropical sun. And in Brazil they have about 80 million cars, each of which is on average lighter and more economical than those in the U.S.; we have at least three times as many vehicles on the road in total. On a sad note, Brazilians have often seen fit to destroy forests just to grow more cane for fuel. This is not a good idea. Forests hold or sequester atmospheric carbon. We need all the trees we can get.

Because of their agricultural situation, Brazilians are able to use their ethanol effectively. Most of their vehicles are fuel-flexible. They have sensors in the fuel system that sense how much alcohol is mixed with the gas, and they adjust the timing of the ignition spark to match the fuel on board. Also, the rubber parts are chosen to tolerate the alcohol, which can dissolve some types of rubber. Without specially specified rubber parts, the ethanol would leak everywhere. That's why, if your car is a 2001 model or older, you are advised not to use fuel with more than 10 percent alcohol, "E10."

Alcohol contains oxygen, so it's a little bit corrosive on engine parts. The special materials, like the rubber, are not exotic technology. The properties of alcohol just have to be taken into account, when the engines are designed and materials specified. Detroit markets flex-fuel vehicles, but in the U.S. the need is not as significant. In the world's third-most populous country, the alcohol to gasoline ratio does not get above 15 percent. The fuel is called "E15." It seems to me that every car powered with an internal combustion engine—and I sure hope there will be fewer and fewer of them—could have the sensor and the rubber parts needed to tolerate alcohol because we can't be exactly sure what the future use of ethanol might be. In Brazil, the fuel can be 100 percent alcohol at certain times of the year, when crop yields and distillery production are high. The problem there can be that water really sticks to molecules of alcohol. Although that's a good feature in a mojito, it's undesirable in a car's fuel tank. If your car engine starts sucking water instead of fuel, well, it doesn't run. Steps have to be taken to get the water out. It adds hassle to the already troublesome nature of internal combustion.

While ethanol has made slow progress in the United States as a stand-alone fuel, it has quietly become a major part of our energy mix anyway, because it is routinely added to gasoline to provide a source of oxygen during combustion. Some of the carbon monoxide in the car exhaust and some of the nitrogen present in the air bonds with the alcohol's oxygen. The extra oxygen at high temperatures makes for somewhat cleaner reaction products in the exhaust. It also dilutes the gas just a little; it cools the flame and delays the combustion of gasoline just a little. So, the gasoline doesn't explode too early before the optimum moment in an internal-combustion engine's cycle. It serves as a so-called anti-knock agent. This is the same function that lead (tetraethyl-lead) performs when it's added to "leaded" gasoline. (Lead is still used in farm vehicles and airplanes. Arrrgghhh . . .)

I am of an age that I remember well when lead was banned from automotive gas. The reason was not the direct effect of lead so much as what that tetraethyl lead did to the catalyst in the once newfangled exhaust-gas catalytic converters. They were introduced in 1975 to catalyze the chemical reaction that makes most of the carbon monoxide ($CO$, not $CO_2$) combine with unburned gasoline to make the good ole $CO_2$. Not breathing lead was, and still is, a huge side benefit. Environmental lead levels dropped, and lead-associated health and developmental problems dropped with them. By the way, catalytic converters were another public-works oriented, top-down regulated technology that many people thought would never work to clean up our air. In fact, they work splendidly.

The people who once denounced catalytic converters as too expensive to put in our cars sound just like so many of our pro-business commentators today who insist that it will cost too much to address climate change. For me, there's an important lesson here. We can be a lot smarter and more capable than a lot of the technology doubters and climate deniers assume. The people who dismiss concerns about global warming seem to be the pessimists who would rather give up than own up to the problems we have all created. The people who worry most about what we are doing to the planet are the optimists who believe we also have the intelligence—we, as a species, working together—to come up with powerful solutions to the problems we're working on that will change the world for the better. Which way of looking at the world is going to produce a Next Greatest Generation? Will it be the ones who give up, or the ones who get going?

Making alcohol from sugarcane and corn is relatively straightforward because of all the sugar that's in there for yeast to metabolize. Varieties of corn in the U.S. carry nicknames like Triplesweet, Butterfruit, and Sugarbuns. I mean just think how much alcohol is in whiskey and rum. Woo-hoo! (I mean, that's just what I've heard . . . about whiskey and rum.) But we can also produce alcohol not just from the starchy sugars in plants like this, but also from the rigid part, the cellulose in other plants. That means we can extract energy from a lot more of the plant, and we don't have to get our ethanol from things (like corn) that could otherwise be used as food.

My father talked very little of his experience as a POW in World War II. But apparently, as the fuel supplies required by the Japanese military were dwindling, my father and his comrades were ordered to produce fuel using the stumps of pine trees. The prisoners would dig large pits, bury the stumps, and cook them to dry them out and make charcoal. Along with that was an effort to capture the escaping gases and reform them to produce liquid fuel that was intended for their enemy's warplanes. The lumber from the trees had been committed to the Japanese war effort long before. The stumps were about all that was left. Apparently, since the process also released carbon monoxide and water, if the gases were captured properly, they could be made into a clear liquid like gasoline or alcohol. It was a lot of work, and the Japanese didn't get much liquid fuel. But it's possible for us, too, to convert the tough cellulose of a plant into liquid fuel, albeit under better conditions.

The big idea is that instead of producing fuel from plant sugars, we can get alcohol from the plant's cellulose. It can be done with the chopped-down stalks and roots, the "stovers," of corn plants. They're mostly cellulose. Right now, the most promising plant is native to the American prairie; it's called switchgrass. It grows very high, nearly 3 meters (8 feet) high. So if you have a field full of mature switchgrass, you have a lot of cellulose on your hands that is a lot easier, than low corn stovers, to chop and grind into something that a modern processing system can use. The switchgrass is ground up and fermented in what's called a digester.

Under the right conditions of temperature, pressure, and a liquid to mobilize it, you can convert the switchgrass into ethanol. Right now, though, the process is not quite economical. It's not clear that it can ever be economical, but it might be a worthy use of farm subsidies. That's for debate. What's not is the energy produced. It takes too much switchgrass and it costs too much to process that cellulose into ethanol. But what if the economics were different? What if the price of petroleum was much higher? Then cellulosic ethanol would appear more attractive. We could even pass a regulation to require anyone or any system that burned fossil-fuel oil to pay for the carbon dioxide release. Then ethanol might be the way to go in certain applications, places where the ethanol fuel is used right there where it's produced.

Along with other eccentric pursuits, I am invested in a company that sought to make diesel and jet fuel from genetically modified algae. It was and is a good idea, but this company has largely set that work aside because they found that they can modify algae to make all kinds of oil, including the organic equivalent of palm oil and oils used in cosmetics. These oils have a much, much higher market value per liter, or cc, or quart, or ounce than jet fuel. So the company is focused on producing those oils. But I cannot help but muse, and scribble rough numbers on the backs of envelopes, about how competitive an algae-based fuel might be if the price of petroleum were a little bit higher. If we all were required to pay a fee (oh, don't call it a tax) on carbon, then gas and oil and jet fuel would cost a little bit more, and companies like the one I'm invested in would be cranking jet fuel through their aquaculture systems. It would be top down; it would be regulation. Wait—in some ways it might be closer to deregulation.

If we were to let the price of gas climb to its natural level, algae-based fuel might be competitive. If you like to muse, consider this as well. The main customer for domestically produced, non-petroleum-based fuel is none other than the U.S. military. Aircraft carriers, which can project military air cover and air support anywhere in the world at virtually a moment's notice, go through enormous amounts of jet fuel. What if the U.S. military made the case that this country has to have renewable jet fuel on the grounds of national security alone? The Department of Defense is funding biofuel research for exactly this reason. If our military leaders decide this is the way to go, it would change things in a hurry.

As I mentioned earlier, there are some good strategic reasons for making such a switch. The U.S. military is deeply concerned about climate change. A tremendous amount of military planning has been done to address conditions around the world as water runs short and populations are displaced, situations that will, almost certainly, result in conflict. If the American economy and American interests were not tied so directly to oil from overseas, U.S. diplomats and politicians could approach the rest of the world in new ways. They could eliminate a lot of military actions that have little benefit other than protecting the oil supply, and they could devote a lot more resources to improving the lives of citizens everywhere affected by climate change—including the citizens right here at home.

Without subsidies, ethanol is impractical from the start. It will not be able to cut us loose from foreign oil, at least not single-handedly. But despite all the complexities that come with ethanol, it looks to me like it might help us make the transition from oil to cleaner fuels, and ultimately to an economy built primarily on carbon-free energy. It might have a role in farm machinery, where it can be produced right near where it is used, and it might be important in cleaning up shipping and aviation. It could be important for the U.S. military, both strategically and technologically. But for most of us, doing our daily commutes, ethanol is not a practical fuel this far from the equator.

Before we leave biofuels in general, though, I cannot help but again point out that the true test of ethanol is to let it compete in a truly fair marketplace. If we create a fee and dividend system in which we all paid a true cost of producing greenhouse gases, then the market can sort out the pros and cons. Then we can truly set our best minds, and our best technologies, free.

# 19

# NASCAR—A CATALYST
# FOR CHANGE

My sister, nephew, nieces, spouses, their exes, kids, and more kids live in and around Danville, Virginia, where it's a big deal to go to the races. The boys of various ages, and occasionally some of the girls and women, will pile into vehicles to go see races in South Boston or other tracks nearby. I've been to the track in Martinsville, which is still the shortest track on the NASCAR (National Stock Car Racing) circuit. It's less than a kilometer long, only half a mile. But it is exciting. The cars go just crazy fast, and they are amazingly loud . . . or LOUD!! But beyond the heart-pounding, are-these-cars-going-to-jump-the-barrier-and-kill-me exciting quality of it, it's depressing—leastways depressing for me as an engineer. Because here I am trying to envision the smart, efficient transportation technology of tomorrow, and there is NASCAR celebrating a very old transportation technology of yesterday. You might call NASCAR the anti-NASA.

As the CEO of The Planetary Society, I'm around space exploration people all the time. So these days when I hear that NASCAR sound, I think, What if NASCAR became more like NASA? After all,

NASA sets up Grand Challenges to inspire companies and individuals to create novel technologies. There are current challenges to come up with ideas to mine an asteroid, build a better space suit, and survive the radiation environment in deep space. Those competitions are kind of like races, with winners who get significant grant money. Or look at Google, which created the Lunar X Prize, a genuine competition to see which private group can land a robot on the Moon and send pictures back home. There's no reason why NASCAR couldn't be like that: a race with rules designed to reward the coolest, most advanced vehicle technologies.

The way NASCAR racecars are designed now, they can go almost 350 km/hour (215 mph) . . . but only around in circles. NASCAR has phased out the "road courses" because the current cars cannot turn very well. They have huge V-8 engines, which are massive and take a lot of energy to speed up and slow down. The engines lift and lower the valves with pushrods. If you went looking out on any U.S. street today, you would have a difficult time finding a single vehicle whose timing cams (egg-shaped wheels) still use pushrods to lift the valves instead of cams driven by a toothed rubber belt (overhead cams). My 1969 Volkswagen Bug had pushrods, and they were recognized as old, 1930s, tech even back then. It takes more energy, and the ignition system has to be downgraded to use pushrods, and so on. But they are traditional, and NASCAR seems to love that, leastways, right now.

NASCAR vehicles also still use carburetors. This is a device that mixes air and petrol or gas before it's put in the engine's cylinders to explode. No modern car is equipped this way. Haven't been for decades. Every modern car uses a fuel-injection system in which fuel is squirted into each cylinder at just the right moment in each cylinder's piston stroke. These systems electronically read and adjust for the outside air temperature, oxygen present, speed that air is flowing in, exhaust temperature on the downstream side, etc. NASCAR's Car of Tomorrow still hasn't come around to this more than forty years after Volkswagen Beetles got fuel injection.

The biggest depressor for me, though, is the fuel consumption or gas mileage. These cars get as little as 80 liters per 100 kilometers, or 3 miles to the gallon (mpg). That is astonishingly bad. By some reports, they sometimes get 50 liters per 100 km (5 mpg). And the whole pack of cars gets better mileage when they are not actually racing, just riding around under the caution or yellow flag. The caution flag has become a more significant part of the activity. There are 30 percent more yellow flags "thrown" today than there were fifteen years ago. It's to bunch the cars up so the outcome is more exciting, with more cars closer together at the finish. The racing is for entertainment, but it begs the question: Why make cars capable of going so fast if you're just going to slow them all down?

Speaking of slowing the race down: Modern NASCAR racers block some of the air available to their engines, albeit at certain venues, at so-called superspeedways, where the straightaways would allow these overpowered, underdesigned machines to go so fast that they become essentially uncontrollable. At these faster tracks, NASCAR requires that racers install "restrictor" plates. These are flat metal plates with four crude holes drilled in them. The diameter of the holes was apparently selected by trial and error. From an aerodynamic or fluid mechanical standpoint, these plates create turbulence in the airflow at the inlet to the engine and increase the already remarkable inefficiencies inherent in the design.

I was at one NASCAR race at the Auto Club Speedway in Fontana, California, during which the lead car had to leave the track briefly and head in for a pit stop. The announcer explained that the car's lug nuts were coming loose. These would be the threaded nuts that fit over the threaded posts or studs that hold the wheel on. The NASCAR promoters want me, as a race fan, to believe that after nearly a century of racing cars, the means by which the wheels are fastened to the cars by expert crews is not good enough to go around in circles under essentially ideal driving conditions. I just didn't believe it. The races are controlled a little bit by the officials, or the racing teams, or somebody. My sister observed how a driver often wins when he (or the occasional she) is in his or her hometown. It reminds me of big-time pro wrestling, which is much more of a show than a competition. NASCAR is about speed, loud, and show. Its vehicles are a little bit primitive, and that just brings me down.

Earlier I mentioned NASCAR's "Car of Tomorrow." It's a fine name, so long as tomorrow means all the improvements we can muster with duct tape in the next twenty-four hours. The Car of Tomorrow is still a standard style of chassis and drivetrain. It's different from previous NASCAR models, but it is absolutely by no means futuristic. It's just another different, inefficient vehicle powered by an engine that is essentially seventy years old. It's nothing at all like what you and I can envision.

Let me offer a backward-looking yet forward-thinking story. When I was a kid, auto racing was where new technology was developed rather than where old technology was preserved or even celebrated. We admired cars that went faster and could be maneuvered well. Next time you're in the Indianapolis Motor Speedway Hall of Fame Museum, check out the STP-Paxton Turbocar. It almost won the 1967 Indianapolis 500 race with a helicopter turbine engine adapted for car racing. The car led for 171 laps, almost the entire race. A ball bearing cracked on the third-to-last lap. Race officials could tell that if they allowed this type of innovative car to come back year after year, racing teams would have a chance to test it further, and make additional improvements to this sure-win innovation. Soon, all the other racecars would be obsolete. They changed the permitted size of the intake duct so that it was too small to enable a turbine car to keep up.

The turbine engine needed extra air to keep it cool. They curtailed innovation by rule change.

I'm not saying we should all be driving turbine engines. I'm saying racing should be about the future rather than the past. I'm suggesting that NASCAR change one rule, just one rule of their detailed rule book (which is not available to the public). How about instead of unlimited fuel, for a 500-mile race, teams be permitted to use only 80 liters (21 gallons), I mean half a barrel, of gas? This would be a rather drastic reduction from the current 100-plus gallons they use today. This sounds radical, and it would be. It would require some innovation. To put this in perspective, if we had a rule today that would cut NASCAR's fuel allowance to even 50 gallons (which is a helluva lot of gas, people), no team could beat you and me in my 2004 Toyota Prius. We could drive this real "stock" (off-the-showroom-floor) car around and around the course for a while. Then we could stop and have pizza. We'd get back in the car and win. No other team could even finish the race, even if we gave them each nearly an oil drum full of gas! Or do it this way: Suppose there was a rule that allowed a team to refuel only once for 800 kilometers (500 miles)—one pit stop for gas. Again, no NASCAR racing vehicle could beat us because they can't go even close to that far. Their cars are just way too inefficient.

As I mentioned at the start of this chapter, I have family members who work blue-collar "working-man" jobs in the South. I've spent a lot of time with them; I love my family. I get it. I understand the appeal of a stock car race. It's just exciting, and I'm all for it. I just want NASCAR to adapt to the new mainstream. I want the circuit to produce vehicles that could compete in races anywhere in the world, and win. I want the racing series to spin off new tech that will do more with less. For me, as an American mechanical engineer, I hope NASCAR decides to look forward rather than backward. How cool would it be, if racecars were all electric vehicles, and there were penalties for using too many Joules or being too loud? Wow, that would be a different twenty-first-century race.

Here's hoping NASCAR officials and teams decide to do some new and cool things rather than the old and slow things. I hope NASCAR gets kids everywhere excited about innovating in automotive design, so that we can go farther on less fuel or even no fuel— just electrons—so that car exhaust is cleaner or nonexistent, so that we can reduce greenhouse-gas emissions and make a better planet for all of us, race fans and regular folk alike.

# 20

# GOT TO GET MOVING ON MOVING

One day, apparently at the direction of my parents, my brother taught me to ride a bike right on the street where we grew up in Washington, D.C. It was scary. I'd fallen on that street a few times trying to learn to operate this statically unstable vehicle. It was rough, literally. The street was a macadam surface: sharp-edged gravel rocks glued to the old pavement below with tar. I had a few scabs already. But wow, did I want to learn. Cycling just looked cool. Well, my brother pushed me around and kept me upright by holding on to the saddle. After a few very short excursions like this, he got me going pretty fast. We were cruising along, and by the time we got to the end of the block, I felt like I was flying. He told me to apply the brake, which I somehow managed to do. I said I almost had the hang of it. He said that I did have the hang of it. He had let go almost immediately after we got rolling. I had been balancing and steering on my own. I couldn't believe it. In barely ten seconds, I had traveled a distance that would have taken my short youthful legs more than a minute to walk.

The efficiency of wheels and other vehicles matter to all of us, not just Young Bill, because almost a third of all the energy we use in the United States goes to transportation, to moving us and our stuff all over the place. We use almost as much energy moving ourselves and our goods around as we use to produce or create those goods in the first place. The transportation sector is a huge one, so even modest improvements in efficiency here can have a tremendous impact. If we want to clean up the world and stabilize our climate, this is an area where we can make huge gains.

Before we get too far down the road here (pun, ouch), consider how efficient a human walking is. Your walking body expends only about 30 watts, and you can walk for several minutes and even hours on end. A world-class bicyclist can crank out 400 watts and keep rolling all day. Now compare that to a car. A modest car produces about 120 kilowatts (150 horsepower). That's 5,000 times as much power as a person walking. You can see why we end up using enormous amounts of energy doing just about anything with an automobile. Try this sometime: Drive your car to a flat parking lot, put it in neutral, and just push it. Now while it's still moving, run around to the other end and try to stop it with your bare hands and the soles of your feet. (This is an excellent exercise to do with your teenage student driver.) The amount of mass we move around just to move around is very large. It takes a great deal of energy to get a motor vehicle moving, to push it against air resistance or drag, and to slow it down.

Since my first spin down that street, I've spent a lot of time on a bike. Every time I swing my leg over the saddle and crank that first pedal stroke, I am amazed at the inherent efficiency brought on by wheels. So the wheels are not the problem with our modern system of transportation; it's the types of fuels and engines we use to make them turn. The net effect of transportation is staggering. In the United States we alone have more than 250 million motor vehicles navigating over 4 million kilometers of paved roadways. Cars not only produce carbon dioxide and smoggy particulates, they spread it all out all over all of our ecosystems. The chemicals we spew as we drive carry enormous environmental impacts.

Getting people to move around less is not a viable option. We've been building up this transportation infrastructure for thousands of years. We're not about to stop using it. Ancient peoples who lived around the Mediterranean Sea—the Romans, Phoenicians, and Greeks—were already traveling over roads using wheels. I've driven a car on the Appian Way, built by the ancient Romans well over 2,300 years ago. You can still see evidence of the Oregon Trail, which ran from Independence, Missouri, to Portland, Oregon in the U.S., when you look down from airplanes and from space. Humankind has paved over 2 percent of the United States. Just consider what that means. Imagine the bottom sheet on your bed was somehow fitted with a stripe of asphalt pavement running right down the middle, like your bed had a concrete or black tar spine. It would suck, or at least be extremely uncomfortable. You'd notice it every night and all day every day for your whole life. Our transportation system is a little bit like that tar stripe; it has already changed the world, and not exactly for the better.

So how can we use that enormous infrastructure more efficiently? I keep coming back to the standard of a person riding a bike. Riding 100 kilometers (just over 60 miles) takes about 1,100 kilojoules of food. That is the energy equivalent of 900 miles per gallon. A bike is 30 times as efficient as a car—without factoring in all the other things, such as materials and energy to create the car or bike and the devastating effects of carbon dioxide emissions. On a bike, our bodies drive our legs, which push the pedals to create a circular motion that is transferred to the rear wheel via a chain. In a car, pistons move up and down and turn a crankshaft, which is in turn connected to the wheels. Either way, the series of parts is generally called the drivetrain.

A quick comparison shows how much the moving parts matter. Bicycle drivetrains are generally well over 90 percent efficient, whereas the very smoothest automobiles top out at around 75 percent.

Getting people to abandon their cars in favor of bicycles sounds like a great idea—I would sure like to live in a world like that—but I recognize that's not a viable option either, at least not in most parts of the world right now. I'm thinking again about my friend Haosheng in China. His family, and many millions of others around the world, are working full-tilt to get their own cars. They don't want to switch to bikes; they want to switch away from them. As I discussed earlier, I think electric vehicles are going to be a huge part of the solution. They provide the kind of personal mobility that people crave, and they have the potential to provide it using clean, renewably generated electricity.

It's also possible that people will find better ways to make use of cars. Buying and maintaining your own car is an expensive proposition. Cars wear out. And when you drive, everybody is in your way, and you in theirs. I can easily envision a future system in which all of our vehicles are autonomous, just like our electric grid and your refrigerator. You'd take out your smartphone and, Uber-style, summon an autonomous get-you-to-work vehicle. The autonomous vehicle company takes care of the car and sends you a bill for each ride. If those vehicles coordinated with each other the way our mobile phones and their communication cells do, and these robot cars were electric . . . oooh! We'd have far fewer car wrecks, less pollution, and more free time.

But if you really want to go for maximum efficiency, you have to look at ways to move people more than one or two at a time. There is no avoiding those two words that seem to drive conservative politicians batty: mass transit. And once you go down that path (another pun), you almost immediately run into two more words that are only a little less polarizing: high-speed trains.

A railroad train does a lot better all around, especially in moving people and freight. (Wait, what else would you move?) If we're a ton of freight, a train is almost four times as efficient as a truck. If we're talking about a passenger train carrying commuters, it can be hundreds or even thousands of times more efficient. Railroads have three kick-ass advantages over cars on roads. First, railroads use metal wheels rolling on metal rails, usually specially alloyed steels. Just ask yourself, which is smoother, a stainless steel countertop or a sidewalk? How about a paved road pocky with potholes? The coefficient of rolling resistance for steel wheels on steel rails is about 0.001. In other words, one-tenth of a percent of the energy needed to pull the railroad car gets converted to waste heat. A passenger car's rolling resistance is often ten times that.

The second, perhaps obvious, advantage of rail trains is that they overwhelmingly run on schedules. You can run a great many trains down and up the same track. Just think if car drivers all got together and took turns, deciding who would use the freeway first, and second, and 4,328th, all coordinated to within a second of time. It would greatly increase the efficiency of almost everything in our transport world. A controlled schedule improves efficiency in another way. Fast as they might go, trains take their time getting up to speed, leastways when compared with cars.

That's the third advantage of a train—acceleration. Our automobiles have to be designed so that they can successfully merge onto our highways. This constrains car manufacturers to build vehicles with engines that are about a third bigger than they would otherwise have to be. Oh sure, we could modify every single highway and traffic intersection in the world so that car engines could be smaller. That's, uh . . . not likely.

From a global energy-use standpoint, trains are just far more efficient than cars, and so they are a lot cleaner, too. In many cultures in many countries, everyone can see that and knows it intuitively. Here in the U.S. there seem to be continual efforts to suppress high-speed trains. In some cases, opposition comes from politicians who simply do not like the idea of government investment in public works. Other times, people object that rail projects just seem to take too long to build. Once again, though, the longest journey by any road, dirt, asphalt, or rail begins with that single step. What if trains were both even more efficient and attractive, sexy in a word?

When you drive the long straight stretches of U.S. highways, like Interstate 5 from Southern to Northern California or Route 70 from Indianapolis to Kansas City, you see trucks tailgating each other. They ride cab-to-tail sometimes. They're doing their best to not lose energy to the air. Unfortunately, close drafting is dangerous. Drivers may not have time to react to trouble ahead. Trains don't have this drafting problem. They suffer a lot less from the effects of air resistance. Most of the resistance happens in the front, where the first locomotive or power car has to push the air ahead aside. When you are driving a car, you can't draft very much. You are constantly pushing hard against the air. If you are conducting a train (and good for you), there are a lot of cars behind the locomotive, and they all basically get a free ride aerodynamically. But what if you could do even better?

If you've ever seen or played with an air hockey table—and I'm guessing you have—you know how good they are at getting rid of friction. The puck slides along on a layer of air. The only significant friction comes from the air molecules the puck runs into as it moves horizontally. Theoretically, you could have a spring-loaded pusher tool that would impart motion to the puck. You could run around to the other end of the table and catch or stop the puck with that same pusher tool now serving as a brake or spring squeezer tool. If the air resistance could somehow be completely removed, you'd get all the energy back. When things are moving horizontally with no friction, you don't need to do any work. Once such a thing is moving, it just keeps moving.

This is the idea behind certain vacuum trains, and especially the much-discussed proposal by Elon Musk for a "hyperloop" train. (Musk, you may recall, is also the mastermind behind Tesla. He's another guy who is big into this "change the world" thing.) Instead of train cars or carriages rolling on train tracks, passengers and freight would travel in a capsule whooshing through an enormous long tube. And here's the trick. The tube would have most of the air sucked out of it. Furthermore, it would have compressors in the front end of the train to suck in the few air molecules that do remain and direct them underneath the cars or carriages to help the cars float above the bottom of the tube. Musk and his engineers claim such a train could travel faster than a jet airliner and not have to break the speed of sound, because there would be so few air molecules around.

The hyperloop idea faces some tall technical hurdles. When I visited the particle accelerator at Los Alamos, New Mexico, I met the technicians there who work extremely hard to pull every last molecule out of their accelerator pipe. It's not that simple. Leaks develop often enough. Pumping all the air out of a train-size tube hundreds of kilometers long may not be such an easy thing to achieve and maintain. The tube would have to remain extremely stable, because any slight shift (even a few millimeters) would disrupt the precise relationship between the capsule and the walls around it. If a capsule sprang a leak, the air would get sucked into the near-vacuum around it; there would need to be elaborate safety systems. And if one capsule broke down or had an accident, the whole tube would have to be shut down until it was fixed, so the system would need to be extremely reliable. These are tough, exciting problems to solve.

Elon Musk's SpaceX is sponsoring the construction of a test track in California and a competition to find executable engineering schemes. We'll see if it all works. It would be a true transportation breakthrough. It could jump-start the whole field of mass transportation. It's worth exploring. Of course, with existing train technology, there is still plenty we can do. So if you're a voter and a taxpayer, and you get a chance to vote for a train instead of a highway, seize it. Go to the polls and change the world a little.

# 21

# MOVING OUR MASSES

As I write, I am living a divided life. Some of the time I am a New Yorker or a Washingtonian relying almost entirely on subway systems that efficiently (if not stylishly) carry one and three-quarters of a billion passengers a year. Imagine arranging for the movement of that many of anything. It's remarkable. And then some of the time, I am a commuter, making my way to the Planetary Society office cursing the traffic on the 101, the 105, the 134, or the 405 freeway, without the energy to wrap my head around megaprojects like a hyperloop train, resigned to a life in which I have to account for all kinds of extra minutes (hours?) of travel time because of car wrecks and traffic delays due to the overcrowding of the roads.

Speaking of roads: What about trucks? Is it such a good idea that trucks use the same rights of way as cars do? The drivers of each type of vehicle are in each other's way. Furthermore, if we can improve the efficiency of car drivetrains, perhaps we could do the same for trucks, buses, and ships. Making fundamental changes to how we use roads may not be easy, but we might all be happier for it. And making all of those engines on trucks and ships cleaner and more efficient might not be such a hard thing, if there were economic incentives. I'll have more on that later, but if a truck or a ship had to pay for the $CO_2$ it puts in the air, the motivation to do more with less would be driving our ships on smoother seas, and our trucks would roll more efficiently even with its squishy rubber tires.

Circling back to mass transit, the success of the New York subway is another prime example of why trains on rails are more efficient than personal vehicles rolling around on rubber tires. Rails have a tenth the friction of rubber tires on pothole-ridden streets. The subway cars are electrically driven so some energy can be recovered with each deceleration. And because most of the lines are underground, where no one and no thing can cross the tracks, trains are not subject to pedestrians crossing against the traffic lights while texting their friends about movie plans. Underground, there are no right-angle intersections to bring a subway train to a halt.

Mass transit is, to me, amazing. It is collective. It's an enormous investment. A society decides that it wants everyone to be able to go anywhere in a logically defined geographic area. It democratizes communication and interaction. Everyone can come and go as he or she pleases, albeit within practical confines. And everyone can choose to use it or not. I know a great many people who don't want any confines of any kind. But as cities fill with more and more of us trying to get around, the confinements just happen. Too many people trying to get to too many places creates conflicts, delays, and losses of productivity (as well as losses of fun).

In Los Angeles it's common to see hard-core fans leave the baseball park in the seventh inning because it will take far too long to get home otherwise. Public transportation is just beginning to get a routine following in that city. That it's taken so many years is evidence of how much of a commitment this kind of public works entails. As we move into a climatically changing future, I strongly believe we will have to enhance our public transportation systems everywhere. Otherwise there just won't be enough energy to go around. Subways are also safer. In New York in 2012, 141 people were hit by trains; 55 of them were killed. Since that system carries so many people, it's not hard to make the case that if all those riders had been automobile drivers, far more of them would be dead. In the U.S., we have more than 30,000 people killed in traffic accidents every year. That's about 100 deaths per million of us. If you just extrapolate blindly, that would suggest about 800 annual deaths for 8 million New Yorkers. Now 55 subway deaths don't look so horrible. And you have to think that each of those was avoidable. With car wrecks and human nature, with fatigue, with alcohol, and with texting, I'm not sure those automobile deaths are nearly as avoidable as the subway demises.

When I mentioned that a passenger's quality of life on a subway train is surprisingly good, my New York–based editor at St. Martin's Press apparently laughed out loud. I'll grant her that automobiles are more private and each person has more room. The driver, though, can't be doing anything else the way a train rider can, and drivers do not get to listen to the occasional self-styled singer or entertaining charlatan begging for money. Train riders read their e-books or listen to their podcasts, or actually talk with someone next to them (I know, I know, very troubling to even consider). But anyway you look at it, trains are safer and more efficient. So I'm a believer in mass transit. From time to time I ask myself, is this a situation where my upbringing as a bleeding-heart progressive induces me to believe in public transportation? Is my analysis objective? Of course, as a scientist, I hope to be objective. There is no way you could handle two billion rides around New York City, or Washington, D.C., or Chicago in cars. You could not move one and a quarter billion riders around London without the Underground. How about three and a third billion riders in Tokyo? Get over it. The trains are inherently better for the planet's ecosystem.

So why doesn't every city have a subway system? In the modern lingo I might say, "because cars." Automobiles and the seemingly endless system of roads and rights-of-way that enable them to go to seemingly any place provide such a high level of service that it's hard for voters and taxpayers to get their heads around the idea that they should pay for another style of transportation.

If you are in a car, keep in mind that everyone who rides a subway or aboveground train is one fewer person on the road in front of you. A taxpayer who wants to drive everywhere has a strong interest in public transportation. Get those other people out of your way. Vote for trains. Not just bullet trains and hyperloops, but the profoundly unsexy and useful light-rail systems. We could change things and make our world so much more efficient if we put our resources to bear on the problem. Let's also keep in mind that all of Bill's-imagined infrastructure is to be created by people who live where the structures are needed. This is to say, you cannot completely outsource railroads. You cannot build a road overseas and put it on or under the ground where you need it. Even if the train carriages and locomotives or power cars are built elsewhere, sooner or later you have to hire local people to install local tracks and controllers. The money for the infrastructure gets spent right where the infrastructure is needed.

As I write, there are a great many conservative lawmakers who feel that any government project, especially one that takes on a big multiyear or even multi-decade task, is inherently bad because it comes from big government. It's a manifestation of The Man suppressing your rights and freedom. As the reader may infer, I just don't see it that way. In the interest of efficiency and quality of life, I feel we will have to have better public transportation. Whether that means nicer train cars, special single-passenger taxicabs, or so much telecommuting that we just don't move around as much in the first place, we must and we will make transportation better.

Many of the advantages afforded by trains also can be said for bicycles. You've probably figured out by now that I am an avowed bike lover, but still. Every bicyclist on the road in front of you when you're driving your car is one fewer driver for you to shake your fist(s) at. People who study the layout of cities are keenly aware of this. When I travel around the U.S. and Canada these days, I see many cities now have public bicycle systems. Riders pay by the ride, or by the day, week, or year. They can ride from one city bicycle station to another. It turns out to be a complex business to determine where to put docks and bikes and how many bikes should be available to citizens in different parts of a city. But with modern computer modeling and algorithms, engineers can manage the system, enabling as many riders as possible to have access to as many rides as they need. We will see in the coming years if these systems expand and become even more commonplace.

When I was fifteen years old, I managed to get a job as a bike mechanic in Arlington, Virginia. I rode my bike about 18 kilometers (11 miles) each way from Washington, D.C. In the muggy Washington summertime, I showed up hot and sweaty, but ready to work. I could do that; I was a greasy bike mechanic. Now as a grown-up (leastways in the eyes of the Internal Revenue Service), I have to show up for business meetings or on camera dressed and smelling like a desk-job professional. It's easy for me to imagine a day when every downtown place of business has a shower and a locker room. I can imagine businesses being created to service the bicycle commuter. A person would come to work, shower, and then put on clean clothes, which had been laundered by a company that serviced many bike commuter–friendly businesses.

Businesses that set up these services could easily be induced to do so with tax incentives. It's another idea so crazy that it just might work. There are a few people out there who have already arranged their lives this way. They commute exclusively by bicycle and take a shower after they get to work. Looking back, I almost did. When I was working on the Science Guy show, Mondays were the writing days. I would bike to work in Seattle's temperate weather and get down to typing. I could see the Big Picture, but I did not have the means to do it full scale. I can imagine, though, that city planners could set up things so virtually any city worker living and conducting the people's business in miserably humid Washington in the summertime could arrange her or his life this way. Well, a cyclist can dream!

It seems like this could work to some extent right away. The system could be expanded with the right tax incentives and planning. In Holland, they say, "Meer fietsen dan mensen." More bikes than people. When you're in Amsterdam or The Hague, there are bikes, bikes, bikes. The roadways are tripartite: There is a car lane, a pedestrian sidewalk, and in between a full-width, comfortable, well-maintained bike lane. All three roads are used all the time, going in both directions. It helps that the weather is generally comfortable. When people get to work after riding at moderate speeds they are clean and comfortable enough to go straight to work at their desks.

In the United States there are a variety of cultural impediments to cycling. If you leave a bicycle unlocked, someone will steal it. People go to great lengths (often of chain) to prevent theft. And on U.S. streets, there are generally insufficient places to suitably lock your bike so that you're not worried about it. As a result, systems of very-difficult-to-steal commuter bikes have recently become popular. A key complaint about these systems is that the cyclists who use them are not especially skilled, and many of them are scofflaws, to boot. They run traffic lights. They turn in front of cars. They endanger pedestrians. To whatever extent that this is true, it is a solvable problem. In the

same way we generally do not allow automobile drivers to run red lights without consequences, it would be straightforward to enforce conventional traffic laws for bicycles. If we ever reach the point(s) in the U.S. and Canada where there are as many cyclists as there are car drivers, the laws and the traditions of enforcement will shift.

Because the Dutch people have much denser, more compact cities, Dutch roadways, bikeways, and sidewalks are well maintained. The bikeways are wide enough, but not too wide. By tradition, only moderate speeds are possible on the compact Dutch city streets, and very few Dutch commuters wear helmets. In the U.S., I do not recommend this practice. We have wide roads and long distances between homes and places of work. So people who want to go fast, go fast. In the U.S., winter weather often breeds lots of potholes. The U.S. road conditions cause cyclist head traumas to be distressingly common and often very serious or lethal.

When I was in college, I was returning from a long ride with my old buddy Steve Fujikawa. We'd gone about 100 kilometers (60 miles). We were feeling good and riding through campus to check out the warm spring day scene (otherwise known as "girls"), and I guess, fundamentally, to get back to our dorm rooms. A sports car, operated by another student, hit me broadside at the intersection of Cornell's Tower Road and East Avenue. Wham! My head hit the pavement and I saw what I describe as sparks; these must be akin to the whimsical stars that swirl around a cartoon character's head after a hard blow to the noggin.

I've mentioned this episode often to my nieces and nephews. I play both parts: "Uncle Bill, were you wearing a helmet?" "No," I respond, "Nobody wore helmets back then." Oh, there were leather head-strap affairs that would provide virtually no protection in a crash. They were derisively called "hairnets." That same year, in 1975, bike helmets with hard shells and crushable styrene foam liners were introduced.

I just missed owning one in time for this crash. These helmets were not readily available; they weren't yet carried in every bike shop as they are today. Later that year and for a few years after, I wore a hockey helmet that had the same sort of liner.

Years later, by the time of my other serious crash, crushable foam helmets had become pretty common. I was wearing one when a driver nudged her way past a stop sign trying to get a look at the lane I was riding in. Her view was blocked by an illegally parked, very large van. I hit her front left quarter panel so hard (how hard was it?) that the top tube of my bike frame was pulled right out of its beautifully brazed Italian socket. I spent some time in midair on my way to the pavement on the far side of the car's hood. I landed . . . on my head, absolutely squarely on my head. But—I was wearing a foam helmet. My head was apparently fine (you can judge that claim for yourself). I jumped up and exclaimed, "My bike! My bike is ruined. . . ."

At any rate, I am a strong believer in bike helmets. Incidentally, I kept that frame in a storage locker until I was asked to host a bicycle safety video for Disney. We showed the frame. I ranted about helmets, doing my best to impress the importance of bike helmets on my young viewers. Then the frame went to a scrap steelyard. For my part, rest assured that I use the public bicycle systems in both New York and Washington—and I wear a helmet. I wear it for two reasons: Those two head-trauma experiences made quite an, um, impression on me, and I'm at a point in my career where I can easily envision the news stories: "Science Man Hit by Taxi—Not Wearing Helmet!" Sometimes I don't wonder which would be worse—the brain damage would be trouble, but that very, very bad publicity . . . ? That might do me in.

When I dream about bike helmets of the future, which is surprisingly often, I imagine a helmet with no chin strap. I'm obviously not sure how this would work exactly, but it would somehow stay on your head even after the first impact with the street or car hood, or what have you. It would need no glue. It would clamp to the back of your head, I guess. It would be well ventilated. I have no trouble recalling my mom and dad going out to a very fancy New Year's Eve party and my father deploying his top hat, which sprang almost instantly from a flat disk to a high cylinder that towered above his forehead. Something like that maybe . . .

I also easily imagine a much more compact bike helmet, one you could somehow fold up and carry in your briefcase. When you needed it, it would spring and pop into the perfect shape. Since making something stiff enough to save your head seems incompatible with something flexible enough to stow in an envelope, I'll keep wearing my conventional helmets and riding carefully and safely. It's just another problem I'd like to address for the sake of cyclists everywhere.

Finally there's the matter of the weather. As in, I'd like to do something about it and not just talk about it. Call me mad, but I think human technology is ready to fix the weather, at least in a small way. Bad weather can certainly be a problem for cyclists. But if you think about it, weather is a problem for any street transportation system. Cars are slowed in inclement weather, as are even the hardest of core bicyclists. I lived in Seattle for decades. It rains all the time, yet no driver is quite used to it. People still skid their cars off of roads. People still rear-end each other because of slick streets and bad visibility through rain-streaked windshields. People riskily swing into traffic because the rain or sleet obscures their views. Bad weather is trouble, coming and going.

So I like to imagine this. Cities are generally built on or near a river. Coastal towns are usually built where a river meets the sea. So generally speaking, cities have bridges. It might not be so hard to convert the main surface, which is nominally open to the elements, into an enclosed tunnel. Cities have popular commuter routes, be it for bikes or cars. Suppose we built tunnels that had tailwinds. I'm not kidding.

Suppose on long straight stretches of bicycle lanes, we constructed tunnels with lanes going in both directions and louvers to direct prevailing winds or fans to force air through the tunnels, so that cyclists going in either direction would be gently assisted by a tailwind. Personally, I think it's a beautiful idea.

The roadways could be built so that the wind circulated. There would be entryways and exits, to be sure, but these would be configured with the wonderful, often subtle, science of fluid mechanics to ensure that there was relatively little energy lost at the entry and exit paths. The walls of the tunnel would be designed so that the air would stick or flow right along the walls and through the main ducts and roadways in the desired directions. The money for the fans would come from tolls or fees. Or stranger still for some of us, the (renewable) energy would be paid for with tax dollars, the same dollars that maintain our city streets. It all comes back to promoting the greater good. Every cyclist on such a commuter route would be one fewer car on the automobile roadways.

One last thing: We as a society have a lot of health troubles that are associated with not getting enough exercise. We have rampant obesity. We have widespread type 2 diabetes. We are out of shape, and it shows. If more people were biking, it's a good bet their health would improve, and our health-care costs would go down. Win, win, win. I know, this sounds like the mad dream of a bicycle-obsessed Science Guy, but . . . just suppose we built one of these tunnels or protected bridges in say, Portland, Oregon. We could see how it went, or how people rolled, and then choose to expand the idea or just turn off the fans and let cyclists deal with the rain as they do today. I'm just a dreamer, but maybe I'm not the only one.

# 22

# RISE OF THE TAXIPOD, ROBOTRUCK, AND BIOPLANE

It is very difficult to predict what the next transportation break-through will be—if there is one. Hyperloop? Maybe. Wind-powered bicycle tunnels? It could happen, possibly. But here's a prediction I feel confident about, the kind I would put money on if I were a betting man. I'd bet you a shiny old hubcap that a hundred years from now, cities will have virtually no motorized personal vehicles. The few that exist will be for people driving out into remote parts of the country-side. Instead, everyone who wants a ride will have a handheld or wrist-wrapped device that will enable that person to hail a cab. The taxicabs will be automated—completely. Same goes for trucks and delivery vans. There will be no drivers.

The automation of the car is already well under way. Just look at the Google car and other self-driving vehicles now in development. I can think of few better places to set that technology loose than on the urban taxi fleet. We can engineer and build driving systems that are more efficient and safer than even our best, most alert, most cour-teous (?) human taxi drivers. These cabs would be self-organizing and self-interacting. They would not bend fenders. They would not cut each other off. Human engineers would design systems to service human riders. We could easily have single-person, or two-person, or four-person taxipods that take each of us where we want to go using far less energy than typical urban taxicab systems use today. They could be all-electric. They could recover a large fraction of the acceleration and deceleration energy by conventionally running the motor in reverse as a generator.

You might at first wonder whether such a system is really feasible, and whether it could be trustworthy and safe. I think the answer is yes. Consider how many elevators you've ridden in your life: probably thousands, maybe even tens of thousands. You've hung dozens or hundreds of meters above the ground in an empty elevator suspended by nothing but wire rope cables. If you work in a building that has an elevator, you trust that system every single day. It's your life we're talking about. Because elevator systems are so mature, we've all come to trust them. They've been in common use for over a century. The most frequent time anyone does get hurt by an elevator system is during maintenance, when a worker has access to an open shaft. It can be a long, long way down. Other than that, accidents are so freakishly rare that they make headlines when they happen.

If we were to embrace a driverless taxipod system, it would take cooperation, technology, and enforcement of a few important and, in a sense, far-reaching regulations. But for me it's so easy to imagine. Instead of a city specifying cabs that look like today's cars and vans, engineers would have a competition. Cities could designate one traffic lane for taxipods, and see if it worked. If these vehicles were a success, the city could expand the system. In other words, it could be done in stages. You wouldn't have to change it all at once. Eventually, the bottom-up design trickle of better taxipods will rule the streets.

I would take out my mobile phone or touch my mobile wristwatch device. The electric pod almost noiselessly glides up. I touch my watch to a pad or a system on board interacts with it wirelessly. I get to where I'm going, and my account is charged the appropriate amount. If a pedestrian or bicyclist accidentally or brazenly crosses my pod's path, the system senses an impending collision and slows or stops in a timely fashion. (You can take an apocalyptic view of such a system and worry that your government will monitor your every move. I've got news for you: Our government can pretty much do that already. For law-abiding people this will not be an issue, because the government of the future like the government of today will have more important things to do than bother with your detour to the sex toy store.) A taxicab system such as I envision here would be based on technology, but it would also be based on trust. We all base a great deal of our lives on trust anyway. What would our roads be like if you couldn't really count on other drivers to stay on their side of the road? If at any time an oncoming car might jump the centerline and come right at you? Yikes! No one would be driving, leastways not anyone who wasn't already a little nuts.

Everything that makes sense about a taxipod makes at least as much sense for a delivery van or truck going into a city. Those vehicles have well-defined destinations, follow well-defined schedules, and often they are allowed only on certain designated roads anyway. They'd be easy to automate. Now imagine that they are also all-electric vehicles. No idling trucks dirtying the city air. No double-parked vans blocking traffic. No crazy lane changes putting you in danger. Daimler recently showed off a partly autonomous 18-wheeler called Inspiration (I guess they got the memo about not choosing names like "Impact"); the technology is coming. It will take many years before robotic trucks replace drivers in cities and on the highways, and the transition will be disruptive, but we need to do it. The upsides are huge. And a healthy, growing, clean economy will easily produce new jobs to replace the ones lost, as has happened over and over in the past.

Okay, here's another big bet I'll place on the future of transportation: The kinds of changes I've been describing for cars, trucks, and trains will apply to travel by air and water as well. The doing-more-with-less revolution will transform all the ways that we move people and goods around the globe. The way we do things right now is just too wasteful, too dirty, and ultimately too harmful to ourselves.

Powered flight was invented in the U.S., and in just over a century it has changed the world. Oh sure, modern air travel may carry occasional inconveniences, but when compared to the time our recent ancestors invested in getting from place to place, the speeds achieved by an Airbus or Boeing plane are almost incredible. We can cross entire continents or vast oceans in a few hours. These are journeys that once took days, weeks, months, or even years. Air travel between the U.S. and U.K. is so commonplace that we now refer to the bitter cold vast North Atlantic as "the pond." The very first commercial transatlantic flight took place just seventy-seven years ago. That's a huge advance in not that much time.

But air travel is headed for trouble if we continue to allow all that $CO_2$, soot, and nitrogen oxides to be blasted into our atmosphere at all altitudes. We are going to have to change something. On top of that, the single biggest consumer of jet fuel, which is refined kerosene or diesel fuel, is the U.S. military. Admirals and generals are deeply concerned about climate change and finding an inexhaustible, reliable source of jet fuel. Running jets on fuel derived from natural gas instead of petroleum is an interesting way to reduce emissions and use more domestic energy supplies. Boeing is running tests on that technology now. But it is only a stopgap solution.

As I mentioned a couple chapters back, a few companies are working hard to produce plant-based fuels, or biofuels, to replace gasoline in cars. If that approach works, it won't just benefit cars. Airplanes could also use biofuels—the tests are going on right now. United

Airlines just ran a test flight powered by fuel from farm waste and animal fat. However things turn out, it seems as though those researchers are on the right track. If oil that we pump and burn came from ancient sea microbes and plants, intuitively it seems like we could accelerate and fine-tune a plant-based oil-production system in the lab and eventually build it on an industrial scale. We could be throwing serious research dollars at the problem rather than relying mostly on start-up companies to come up with the organism or the process. Agriculture companies spend millions of dollars a day working to perfect crops like soybeans. It may take a hundred times that to perfect plants that produce competitively priced jet fuel from plants.

While people work on that, the bigger opportunities in air travel are probably not in the fuel; they are in the planes themselves—specifically, what they're made of. Even if you haven't seen the movie, you're probably familiar with the famous scene in *The Graduate* in which Mr. Robinson (played by Murray Hamilton) pulls aside Ben Braddock (played by Dustin Hoffman) and whispers, "Plastics. There's a great future in plastics." For now, airplane wings are made of shiny aluminum panels held tightly together by a great many rivets. When the flaps extend, you can see lots of metal torque tubes and linkages inside. Imagine all that hardware being made of plastic, technically called "composites." They're here now. Plastics are lighter than metal, and they can be stronger as well. By one industry estimate, a composite-based airplane could reduce fuel use by 15 percent, which is an excellent way to do more with less.

Plastics and taxipod style automation could finally bring to reality that icon of old-school visions of the future: the flying car. There are some pretty good reasons why flying cars never happened. Picture all of your neighbors buzzing around the neighborhood, along with a few local teens, plus that old guy who always has his turn signal on. How long could any of us survive in that world? Nevertheless, many companies are engaged in building personal airplanes for trips at higher speeds than you can manage in a car. They have names like Terrafugia and Skycar. One reason these companies are even considering getting in this business is that high-performance, high-strength, low-weight materials are now available. These personal aircraft, if they do come into commercial existence, will be able to fly because they are very lightweight. They'll be made of plastic composite materials, and the metal in the engines will be aluminum, magnesium, and titanium. And you better believe they will be partially or totally self-piloting. That is the only plausible way to deal with the horrific mess of setting us loose up above cities behind joysticks and rudder pedals.

Hybrid-electric and all-electric power is coming to airplanes, too. Sounds hard to believe at first, if you are still thinking about how much old-school batteries weigh. But battery technology is advancing rapidly, and engineers are coming up with some very clever ways to put it to use. Airbus, the European aerospace giant, is working on a battery-powered two- or four-seat airplane called the E-Fan that runs on lithium-ion batteries. The company recently flew a prototype across the English Channel, and promises a production version by 2018. Airbus and other aerospace companies are also exploring larger passenger planes that would use a mix of battery and petroleum-based power. The idea, as in hybrid cars, is to use batteries to squeeze more miles out of each gallon of liquid fuel. Commercial planes will probably become mostly or totally automated, once again in the name of safety and efficiency. Get used to seeing a lot fewer drivers and pilots and more software engineers in all aspects of your life.

Beyond plastics, there are carbon-fiber composite materials. This stuff is akin to fiberglass, but the fibers are carbon rather than silicon dioxide (glass), and they are oriented in the exact direction needed to react to the stresses involved, or carry the loads needed. These aircraft are stiffer, stronger, and lighter weight than aluminum and titanium.

And beyond today's carbon fiber, I expect we will see materials constructed from carbon nanotubes: single carbon molecules that are meters or even kilometers long. This material has the potential to be 1/6th as heavy as steel, but 10,000 times stronger. Airplanes could be made fantastically light compared with what we have today. The potential efficiency improvements boggle my mind.

There is a final piece of the transportation puzzle that is ripe for revolution, and that is international shipping. Look at what you're wearing right now. Chances are it came from overseas. How about your dishes and flatware? What about you car or bike? A huge fraction of our everyday consumer goods and industrial equipment is manufactured elsewhere and shipped here. Modern container ships are huge, and right now, they are allowed to pump a great deal of carbon dioxide and other pollutants into the air we all share. The International Maritime Organization has set improved efficiency standards for ships built after 2013, but that is only a start.

Ships at sea present a great many opportunities to improve their propulsion systems, clean up the dirty fuel oil they burn, and reduce their emissions. We can motivate these changes with reasonable, enforceable regulations. We could transform shipping from a dirty business to a clean one using the technologies for clean diesels and perhaps even electrical propulsion that we almost have right now. It would change the world in about forty thousand ways. That's the rough number of container ships plying the ocean even as you read. Let's do more with less on land, sea, and in the air. As we say at the travel agency, "Let's go!"

# 23

# THE WATER-ENERGY CONNECTION

As an unskilled surfer, I can tell you that one mouthful of seawater is tolerable, but three or four or five mouthfuls is not. Not that I would know, of course; that's just something I've heard. And should you ever find yourself adrift in a lifeboat or raft at sea, freshwater is a big concern, because even though you're surrounded by water, a few sips of the sea would make you sick. Although our bodies need a little bit of salt, you and I can't handle the concentration of salt present in almost all of the water on Earth. For us, ocean water does not do the job. You and I, along with dry-land green plants, need clear, fresh, non-salty water to survive. You could choke on the irony. In California we have a state bigger than many countries, where people are out on the beach, surfing the waves elbow to elbow, and a kilometer away their lawns, gardens, and neighbors are facing a severe drought.

The issues of water, energy, and climate are all linked. Humans use a lot of energy to acquire and move water. Right now, that energy produces a lot of greenhouse gases . . . which contribute to a climate change . . . which disrupts the supply of freshwater . . . and off we go again. Salt-free water comes from rain and melting snow, and in many parts of the world there is just not enough of either. It's a problem that is likely to worsen in California and many other places due to changing weather patterns and a warming world. Meanwhile, rivers are increasingly tapped out, aquifers are dropping, and the population is growing. Soon the world is expected to have about 9 billion of us. That's a lot of people who will need to drink water, wash with water, and eat water-consuming crops. We've got to find a way to provide freshwater to everyone in some new way.

You may not have thought about it, but just about every single person on Earth lives near a river or stream. That is no coincidence; it is the way that human settlements have developed for thousands of years. Even if you live on the ocean shore, somewhere nearby is a river of non-salty water. All that salt-less water comes from the sky. Just consider a medium-size cloud. It doesn't cover the whole sky, but you notice it. It might be 20 kilometers across and a kilometer thick. This is the size of cloud that engenders some doubt about your picnic, but you pack the basket anyway. Such a cloud contains at least 100,000 tons of water. If it's an overcast day with clouds from horizon to horizon, you're looking up at millions and millions of tons. That's how animals like you and me, along with flowers, birds, and trees, can get enough water to live.

The Sun drives water from the surface of lakes, ponds, and seas high into the sky, where the water condenses to tiny cloud droplets. Clouds can form, and remain high above, because water molecules weigh less than the nitrogen and oxygen molecules that dominate our air. Water just floats on up until the molecules spread out enough, and cool off enough, to turn into tiny droplets of liquid. As those encounter each other, droplets become drops. Then they drip on an atmospheric scale, falling on us by the ton. And that snow and rainwater contains no salt. When the water changes from a liquid to a vapor it leaves the salt behind. This process of evaporation, condensation, precipitation, and collection of water is what we call the water cycle. It's what sustains us all, and it starts with the process generally called "distillation," from the Latin word *stilla*, for "drop" or "drip."

When it comes to distilling, you can't beat nature. A great many people visit Las Vegas, Nevada, each year. If you're one of them, I encourage you to mosey out by tour bus or car to see Hoover Dam. It's an amazing feat of engineering. But to me, even more amazing is the huge reservoir "impounded" on the upstream side of the dam. Lake Mead holds about 30 cubic kilometers (24 million acre-feet) of crystal clear blue water. Or should you find yourself near The Dalles, Oregon, consider a visit to one of the huge dams there. I love good ole Electron, which is nestled below the enormous expanse of the Grand Coulee Dam. This one is not as high as Hoover Dam, but it is much, much longer, 1,600 meters (almost a mile) across. It impounds Roosevelt and Banks lakes—hugely, wildly impressive.

Water projects like these have reshaped the world. Local ecosystems are transformed when nearly limitless volumes of freshwater show up. As a rider in the Cannonball 300 bicycle competition (Seattle to Spokane, Washington: 300 miles in one day, Finished 1st, "unsupported"), I've ridden my bicycle through areas fed by the Grand Coulee water project. You become aware of the water, the lakes, ponds, or irrigated farm fields an hour before you see them, just by their smell. Water nourishes plants, which in turn produce pollen and all sorts of other fragrances to attract pollinators like birds and bees—and ultimately us. It's water distillation and collection on a grand scale.

At my local drug store, a bottle of distilled water, four liters (a gallon), costs about three dollars. It says on the label that it's been "steam distilled," or "flash distilled." A water-supply company gets water hot enough to evaporate, then lets the steam come in contact with a cold surface. Drops form and drip into a collector. It's a brute-force way of doing what nature does, only a lot more quickly. If you're buying only one bottle, distilled water is about the same price as a fancy cup of coffee. If you're getting ready to do the dishes or flush a toilet, it's a very expensive means to get pure water. Distillation works, but it takes a lot of energy to run the process. But because of its value in all sorts of industrial processes, as well as how well it works in clothes irons and steamers at home, distilled water is fairly common in the developed world all the same.

One day, the U.S. Navy allowed my Science Guy television crew and me onto the USS *Ohio*. It's a Trident-Class submarine, huge, about 170 meters long (560 feet). If it were an apartment building, it could support a block-full of businesses: a pharmacy, a hardware store, laundry and dry cleaner, and a grocery store. But it's a ship. It floats and sinks on and in water. Ships like this distill all the fresh, non-salty, water they need because energy is not much of an issue for the crew of one of these vessels. They have a nuclear reactor power plant on board. They do not need to go into port or meet up with another ship at sea for fuel—for years. They get all the freshwater they need by boiling a bit of the ocean. They even use electricity to make their own air, by separating the $H_2O$ of seawater into hydrogen and oxygen. Purifying water by distillation is the oldest trick in nature's book, and one of the oldest in ours, too.

Like you and me, most people do not live on a nuclear-powered submarine. (Perhaps a few of you are reading along under the waves; I hope so.) Unlike being aboard a powerful submarine, the energy required to distill water is generally a very big consideration, so engineers have come up with a couple of important tricks to keep the energy costs down. When anything changes from a liquid to a gas or the other way around, we say it's undergone a "phase change." When it comes to getting water to change phase, to evaporate, a surprising amount of energy has to be exchanged. We can get it very hot, but that really sucks up a lot of energy. Alternately, we can make the pressure above the surface of the liquid water low. With fewer molecules of air pressing down on the liquid water surface, more water molecules can escape as the gas we call water vapor. In many distilling systems, there are vacuum pumps to lower the pressure in the distilling vessel or chamber. It's part of the flash in what we call "flash distillation." These pumps use less energy but they still use some, which has to be factored into the overall efficiency and cost of the system.

Here's another trick that I find elegant. If we place our flash distiller way down in the ocean, say at 50 meters (150 feet) deep, we can have it communicate with the surface through a pipe, like a snorkel. Then we let the seawater into the distiller at this depth through a throttling valve. It immediately experiences the lower pressure of the surface, so with a relatively small amount of heat, we can boil or "flash" water down there. The only ancillary energy cost is pumping it from down deep to the surface. It's a clever way to put physics to work, but we have to have all this plumbing and a heater deep enough and close enough to shore to make it worthwhile.

On a cruise ship, they have the same problem as on a submarine: Water, water everywhere, nor any drop to drink. Modern ships don't use distillation at all, though, because there's a better technique that doesn't require nearly as much energy. Instead, the ship's systems force salty water through a membrane that lets water molecules through but holds back the salt. You can try this for yourself, if you're inclined. It's an easy but important science demonstration that I hope everyone tries at least once. Soak two raw eggs in vinegar for several hours, or overnight. The shells will dissolve into the vinegar and disappear. (If some bits of shell remain, peel them off ever so gently.) You're left with naked eggs that have no shells, but are still held together by thin skins or membranes. Now put one egg in distilled water. Put another egg in a solution of very, very salty water. You can just add salt and stir until no more salt dissolves. After about a day you'll see that the egg in distilled water got bigger while the egg in salty water got smaller. In each case, the water is moving through the egg's thin membrane from the less salty side to the more salty side. Bird eggs contain a bit of salt. When surrounded by less salty water, the water in the salty situation works its way into the egg. When surrounded by super-salty water, the water in the egg works its way out. In other words, water moves through the egg's membrane but the salt does not, leastways not easily. This is the process we call osmosis. The word comes from the Greek, "to push."

To create drinking water, we can drive the osmosis process the other way by forcing the water backward through the membrane. By pressurizing the salty side, we can produce freshwater on the non-salty side. We logically call this "reverse osmosis." With a durable membrane, we can do this on an industrial scale. A modern cruise ship produces about a thousand tons of freshwater every day. The rejected salt gets tossed overboard. Modern ships use a multilayered arrangement of plastics often called a thin-film membrane. It's a sieve for molecules, held together by a strong backing layer. The pumps often drive the system around 6 megapascals, or 900 psi. That's 60 times normal atmospheric pressure, and about 30 times the pressure in your car's tires.

Flowing through filters takes less energy than forcing a phase change, so reverse-osmosis systems require less than a third of the energy of boiling water or flash-distillation systems. The process is good enough that it is used to supply drinking water to dozens of pretty big, dry cities around the world, including Yanbu in Saudi Arabia, Adelaide in Australia, and Carlsbad in California. If you have a good-enough filter to keep tiny particles out of the membrane, and you have a system of pumps strong enough, you can produce millions of tons of happy drinkable water a week. The municipality still has to maintain and replace the reverse-osmosis membranes and clean the filters.

But using less than 1/3rd the energy compared with flash distilling makes it a must-have, if you're an animal or plant around there bent on staying alive. In the present world, reverse osmosis is a good-enough technology.

In the future world, we will need to do better. We are going to need more water, and we are going to need to get it using less energy. How are we going to do that? Well, rain clouds provided the inspiration for distillation. Eggs provided inspiration for osmosis. Perhaps nature has some even better example to offer us? Indeed it does. Follow me into the swamp and you'll see.

# 24

# TIME TO GET THE SALT OUT

Doing the Science Guy show, I found myself wading through some swamps—literally, and on purpose. It is no trouble at all to get lost there. Every tree, berm, and slow-moving flow of water looks oh so very much the same. Furthermore, you can't easily tell which way is downhill. If you're in an estuary, which is generally where you'll find these wetlands, the tides ebb and flow at confusing times because of all the vegetation. The movements of the water are so subtle that you typically can't tell if it's going in or out. There is an enormous diversity and abundance of life in those places, which can also get confusing. And it's easy to get distracted just marveling at the living things there, many of which are just extraordinary. To survive in a swamp, they have to perform elaborate feats of chemical engineering, specifically including one that humans will be needing soon to deal with our warming world.

Should you find yourself paddling about in the Everglades, take a moment and examine the leaves of a mangrove tree and you'll see what I mean. The leaves are often covered with a white film of salt crystals. Mangroves thrive in places where the freshwater flow from the north rolls up against the salty Atlantic Ocean. These trees can live in salty, salty water. Some species have roots fitted with membranes so fine that salt from the sea can't get in. They allow only water molecules to pass, enabling the trees to drink seawater. Even stranger to me are the mangroves that have glands in their leaves that reject salt. They use some of the energy they get from the Sun to drive the salt out of their leaves and leave it to dry—those salty crystals.

Just think of the possibilities if we could mimic what the mangroves do, if we could find an organic way to desalinate on an industrial scale. Right now, that ability is beyond us, but we know it is feasible; we see the mangroves. What if we created crops or shoreline desalination plants that could turn seawater into freshwater for us, on an industrial scale? Those plants might save coastal cities from drought. They might pump water to inland farms, keeping crops green even as temperatures rise and rainfall patterns change. It could hugely improve on what we can do with reverse osmosis. Mangrove-style water-purification plants could bring clean drinking water to parts of the world where it is scarce or unavailable. In short, the global standard of living could keep improving, unstoppable.

I mention mangroves because I find them so striking. But they are not the only species that can do this. There are seabirds with special glands above their eyes that do the same thing. They use a dense network of blood vessels and membranes, not unlike the membranes inside eggs, to push salt from their blood. Salt crystals emerge right out of their noses . . . or rather, out of their nares, the bird equivalent of our nostrils. By comparison, the desalination techniques that humans use right now seem heavy-handed. Surely we can take another lesson from nature and find a better, more efficient way to get the salt out. We just need the magical material that separates water from salt as readily as the mangroves and the seabirds do.

Well, we may have just found that material, and better yet it is just a specialized form of a very common element—our beloved carbon. The most familiar type of carbon is the kind you see in charcoal, or in the tip of a lead pencil (which does not actually contain any lead, and hasn't since ancient Roman times). That black pigment is carbon in its natural form, graphite. Carbon atoms have four chemical bonding sites. In nature, carbon atoms link together using just three of their available bonds. Graphite forms layers that geometrically resemble honeycombs, six-sided hexagons linked corner to corner. The fourth carbon bond is an electron that's bound strongly enough to remain near the carbon atom, but loosely enough to fly above or below the hexagonal pattern. That electron repels the next layer of hexagons above or below, causing the graphite layers to slide around quite easily. That's why graphite dust makes a good lubricant.

When I was in eleventh grade, I suggested to my chemistry teacher Ms. Hrushka (and the class) that we could perhaps organize the graphite layers into large sheets to control a pencil's hardness. In a way, I had stumbled on the general idea. Over the past decade, researchers have been working on ways to fabricate and isolate single large sheets of graphite. We're talking about one-atomic-layer-thick sheets of carbon hexagons. Researchers call it graphene, and it is amazing stuff.

Today's biggest sheets of graphene are about the size of a floor tile. Even though they are solid carbon, you can see through them because they are so thin. Single graphene layers look like thin gray sheets of flexible plastic. The floating electron is bound to oxygen, and it's bound tight into a hexagonal pattern, making the material exceptionally tough. Picture a honeycomb or a chain-link fence in which each of the links is made of hooked-together carbon atoms and is less than one billionth of a meter wide. That's what graphene looks like when you examine it under an electron microscope. Any applied force gets distributed throughout the pattern. And get this: When graphene sheets are put under stress, they are about 1,000 times stronger than steel. And yet a sheet large enough to cover a football field would weigh only about 1 gram—less than a twentieth of an ounce. Wow!

Along with all those other extreme properties, graphene can do something really useful for making clean water. Take a look at a glass of water and gently spin it on a tabletop. You'll see that the water lags behind the glass at first, but then it starts to spin, too. If you have trouble seeing the spin, just sprinkle a few flakes of pepper on the surface. At this point you might remark, "Bill, dude . . . of course the water spins with the glass. . . ." But why is that? Why doesn't the water just slip right on by the glass surface?

Water, like a great many fluids, sticks to a great many surfaces. Think about oil and metal, vegetable oil and wooden bowls, and of course paint and just about anything. A water molecule gets a little spin from each glass molecule it rubs against. As you spin the glass the molecules tumble and the small body of water starts spinning. In a one-molecule-thick layer of graphene, a water molecule slips through without getting that repeated twisting and tumbling. It's out on the other side of the sheet before a second carbon atom can give it a tumble. Chemists refer to its "slip length" being longer than the graphene is thick.

Imagine pumping water through a large, completely uniform sheet of graphene. The molecules of water never get tumbled; they never stick. They pop right out on the other side of the membrane with very little friction, and the salt molecules get left behind. Such a superthin extensive sheet is, so far, quite difficult to fabricate. Researchers have come close by having methane gas carry carbon atoms up against a fancy polymer or plastic backing sheet or plate. The gas molecules move right on through specially prepared perforations in the backing, but the carbon atoms stay stuck and naturally link up in the honeycomb pattern of graphene.

Right now, these membranes are laboratory-only materials. They're grown ever so carefully. But researchers are by some accounts very close to producing graphene on industrial scales. If this works out, we may be able to desalinate water at a fraction of the cost of reverse osmosis, which is in turn cheaper than flash distillation. Just imagine what this technology would do for the world. If people everywhere had access to clean water, people everywhere could avoid a great many waterborne diseases; they could raise enough food for their families; they could have the quality of life that all of us in the developed world take for granted. The water would come from the ocean, and sooner or later it would end up back in the sea. Unlike pumping groundwater, we would no longer be drawing down a finite resource. The world has had the same amount of water for billions of years. If we managed things properly, we could desalinate a great deal of water, make use of it, and return it to the ecosystem without severe environmental impacts. We could soften the effects of drought in a warming world, and we could grow crops in places that are now (or will soon be) too dry.

Speaking of water and affecting the environment, people have suggested that we pump water over the same vast distances that we pump oil and natural gas. My feeling is, why not? This would be a kind of geoengineering. It is different than the kind of deliberate global changes I described in Chapter 6 . . . except that it really isn't. Humans already do all kinds of things that change Earth on a regional and global scale. Next time you find yourself flying over California, look down and see if you can spot the California Aqueduct and the Colorado River Aqueduct. These are enormous human-built rivers, perfectly engineered to keep water flowing downhill for hundreds of kilometers. With these systems in place, we can irrigate enormous expanses of farm fields and water a great many lawns in Southern California. We have changed the semiarid desert into an extremely productive farming region. Now we need a way to keep the water flowing, even as the climate is changing.

Right now, California is in a deadly drought. Along with killing crops in the fields, the drought has contributed to a large number of wildfires. It is a pattern that we can expect to see more of in the future. Much of North America relies on produce from the fertile valleys of California. According to climate computer models, California will receive less and less rainfall over the coming century. Furthermore, and almost as important, northern Californians resent sending rain that falls where they are to people in the sunny south just so their fellow citizens can, in a sense, squander that water on lawns and in swimming pools. It begs a series of connected questions: Should we ration water? Should people be allowed to have pools and yards? Should we subsidize water for farms? Should we abandon California farms in favor of agriculture elsewhere, even though we have invested heavily in systems to enable growing and distributing all that food?

The questions would be tough ones even if they applied only to California. In truth, they are global questions, and in many parts of the world the issue is not swimming pools and lawns but a basic shortage of water for drinking and growing essential crops. Global climate models forecast that the United States will be hard hit, but so will Europe, the Middle East, northern Africa, southern Africa, and the northern part of South America. According to the National Science Foundation, we are facing relentless thirst. But there must be a way for humankind to get through this, to make the adjustments necessary and come up with new sources of water. Cutting back alone just won't cut it. In coming decades the world will have more people expecting a lifestyle that is better than what we are enjoying today.

What if this graphene technology proves indeed to be an economical way to desalinate water from a nearby ocean coast? I admit, it's the scale of the idea that's daunting. We're talking about sloshing billions of tons of water around a state bigger than a great many of the countries in the world, every day. The keys will be the graphene desalination filter, and a renewable source of electricity to drive the pumps. Clean electrical energy is crucial to pull out of our downward spiral of climate change and drought.

It seems reasonable that we could use graphene to make desalination a viable alternative to diverting rivers and draining aquifers. We could run the desalination plants with clean wind energy in the morning and evening, when the wind blows most vigorously, and with solar energy in the middle of the day. Then at night, the pumps would rest. Perhaps we will be able to use that nightly downtime to maintain things, replace some gaskets, scrub a filter or two. By the way, notice that when it comes to water, we already store it in huge reservoirs. In the short and perhaps medium terms, we could desalinate and pump only at times when renewable electricity is available and leave the water in reserve for other times. That would alleviate some of the irregular nature of wind and solar power.

This scheme would free up our existing power plants to provide the baseline electricity we use at night in our homes. Over time, we could roll out enormous energy storage systems (batteries, pistons, or pumped water—or all of the above) so that we could rely almost entirely on renewable energy. We might use small nuclear plants to produce a new lower baseline of energy, when the renewable systems are resting. We would not need fossil fuels at all. At the same time, we would want to develop ways to recycle water and use it more effectively, so that each drop goes as far as possible. It is a vision that is simple to describe but complex to create. If we can economically desalinate, the remaining technical obstacles are workable. We can do it, and if we do we'll be unstoppable.

# 25

# FEEDING THE WORLD

I don't know you, but I'm sure that at some point in your life you've been hungry. If you're of a certain economic status, you probably know hunger better than someone like me will ever know. Making sure that everyone around the world has enough to eat is one of the most important things we can do in watching out for ourselves—our species, humanity as a whole. Growing and distributing food requires a lot of energy. Climate change poses a new set of challenges for many of the planet's most productive crop regions. And we are chasing a moving target: Since there are more of us every day, the need to do more with less is starkly apparent.

Think of it this way: Farmers use solar energy to feed us all, and "they" aren't making any more of that, but "we" sure are making more of "us." Right now, the human population is just over 7.3 billion. By the year 2050, demographic researchers project that there will be about 9 billion people alive. Earth isn't getting any bigger, and the Sun isn't going to put out any more energy, so we need to intensify our agriculture, but in a sustainable way. Simply feeding everybody in the year 2050 cannot be the single goal. We have to feed everybody in 2050 while also ensuring that we can continue to feed everyone long, long after that. We have to tread more lightly on the planet.

Quick quiz (what, nobody told you?): What human activity has the greatest impact on the planet? Driving, perhaps, or transportation? No. What about construction—building cities? What about heavy industry, mining, and manufacturing? Nope and nope. If we add it all up, the economic sector that uses the most of Earth's resources and produces the largest environmental change is our agriculture. Our farms produce greater volumes of more greenhouse gases than all of our cars, trucks, trains, ships, and airplanes combined. Meanwhile the global population is growing, headed past the 9 billion expected in 2050 to perhaps 10 billion by the end of the century. If we want to feed everyone—and ideally feed them better than they are getting fed today—we're going to have to think big. Here are a few numbers that frame the task ahead.

The total area of the sphere that is Earth is about 510.1 million square kilometers (196.6 million sq. miles). Most of that, 71 percent is covered with ocean. That leaves less than 30 percent for all of us land plants and animals. Humans, our one species alone, farm about 11 percent of that dry land area. Even with all that, about 1 in 7 people is hungry. It's not that we don't have enough food, not exactly. Experts estimate that only about 55 percent of the calories we produce on farms actually gets successfully incorporated into a human's diet. The other 45 percent is lost to pests, plant diseases, spoilage, and especially waste. In the U.S., more than one hundred tons of food are wasted in homes and restaurants every *hour*. Shameful. In the developing world, tons of food are lost to the lack of refrigeration and the inability to successfully transport it from where it's grown to where it would be eaten.

Too much of the food we produce doesn't get to everyone who

needs it, at least not in a timely fashion. Although this seems like a solvable problem, we have to get to work. We need to waste less. We need to preserve the food we produce. We need to distribute food more effectively. And at the same time, we need to produce more food in total. This is one of the defining challenges for the Next Great Generation. I feel that getting more good food to all of us will require two enormous public works–style advancements: providing electricity and clean water everywhere these utilities are needed.

As I mention from time to time, my uncle Bud was an explosives salesman. He was also a gentleman farmer. It was big fun when he assigned my cousin and me to run the tractor. But if you asked him or any of his farmer neighbors what was on his mind, they wouldn't answer right away because they were looking at the sky and trying to assess if it was going to rain. Farmers think about the weather constantly. Their livelihood depends so much on how much rain falls and when it falls. Farmers focus on water. If you ever have the occasion to fly over the western U.S. or Canada, look down. You'll see enormous green circles. These are crops irrigated with enormous machines that roll around that circle from a central pivot and spray water evenly on the round field. Other fields are perfect rectangles. A crop grows only where the water goes.

In California during the summer of 2015, as in many areas of the world, water is becoming increasingly scarce because rainfall patterns are changing, a trend that will worsen in a warming climate. In some cases, the need for water begets violence. Disputes over water rights and control of dams has helped fuel conflict in the Middle East. At the political level, nearly every part of the world is running into clashes over who controls the rights to rivers such as the Colorado River, whose water is coveted by multiple states in the United States, leaving barely a trickle for Mexico to drink. There are also a growing number of quarrels over rights to underground water. You may not have thought about it too much. But just as there are river systems and watersheds above ground, where we can see them and paddle canoes on them, there are also extensive river systems underground. There is a lot of water down there. Of course, there used to be a lot more, until humans showed up with the technology to extract it.

I worked in the oil fields of west Texas and New Mexico for a couple years. When we produced oil, we also extracted oily, salty water. Down deep in the ground with the ancient sea life that decayed into oil, tar, and natural gas, there is a lot of water. To keep the subterranean pressures up and the wells flowing, that water is generally pumped back down. It comes from an aquifer, water-bearing rock. In some areas, some of that water is actually pretty good. It is well suited to irrigating crops and pastureland for livestock. People fight about it. Lawsuits are raging right now. Wells pumping up water in Texas affect wells in New Mexico. It's just like people upstream in a river using all the water before people downstream can get a sip. Pumping one place can dry up another place.

Exploitation of groundwater is a growing problem in the Middle East. Instead of relying on natural flows from mountains through underground aquifers to the villages in the lower lands, people are unsustainably pumping water up from the deep aquifers to irrigate their crops at higher elevations. Note that it would be a difficult business to supply water at higher elevations naturally, because the water flows underground in those areas. It's modern pumping that makes this sort of thing possible in the Tigris and Euphrates region as well as in Texas and New Mexico. It's disrupting agriculture, and it has the potential to disrupt . . . well, everything. What to do?

As you know if you've read this far, I hold high hope for an industrial-scale way to desalinate seawater. But when it comes to irrigating large tracts of farmland that lie far inland, this is not such an easy thing to accomplish. Remember that old saying? There are two ways to be rich: You can have more, or you can need less. In a warming world, much of the future of farming may depend on finding ways to grow more food with less rainfall. What if crops that were nutritious and delicious could grow with less water? What if we could grow the food we need today and in the coming years on less land? What if we needed less fossil fuel to run our farms? What if we could raise good crops with less farm machinery needed to plant, cultivate, and gather those crops? Well, these are the ideas driving modern farmers and agricultural companies.

In my college lectures, I enjoy showing slides depicting precision farming. It is now routine for farmers in the developed world to assay the chemistry and microbe diversity in their fields' soil to a resolution of less than 10 centimeters (4 inches). The data are gathered by satellites in Earth orbit. Using information beamed down from space, a tractor planting or tilling or fertilizing or treating the farm field can be synchronized so that each individual crop plant gets its nutrients optimized. Each plant gets just the right amount of plant food or herbicide. This is an extremely powerful way to do more with less.

When I travel in Delaware to visit my family, as I do every year, I marvel at the farm-irrigation systems. They're huge. When I go to a baseball game in a modern professional stadium, I cannot help but revel in how beautiful the green manicured field is. Nowadays, many irrigation systems are hidden. They provide more water to the grass with less lost to evaporation and spillage from overspraying. I compare them to the systems I use to water my garden and I wonder if there isn't a big opportunity. I will not be surprised when systems for huge farms are huge and small at the same time. I've got a feeling a modern irrigation system will be developed that enables water to be delivered to a huge field with less energy to drive the pumps and less water lost to the sky.

In my own garden, on a small scale, I've experimented with

hydroscopic beads: small gelatinous spheres that hold water. When the soil is wet, they absorb water like spherical sponges. Later, they slow the drying of the soil, and they are surprisingly durable. Soon I expect this hydroscopic technology will be employed on a huge, industrial water-saving scale. Suppose the beads retain just 5 percent of the water that would otherwise evaporate. (Certain manufacturers claim 10 times that.) That 5 percent would represent billions of tons of water every week in virtually any farming area anywhere in the world. If this technology can be developed and distributed, farmers everywhere would be doing more with less.

Taking this idea one step further, I imagine widespread systems for storing electricity on farms. With all that open area, it would be easy to set aside a little bit of land for a solar-electric system or a wind-turbine system sized or tuned just for that single farm or a group of farms. This literal kind of "wind farming" is already happening in parts of the American heartland. Farm machinery of the future could be developed to run on electricity without the need for farmers to purchase and burn the billions of tons of fuel they currently do every year. We would have to encourage the development of electric farm equipment. We could make it profitable by charging a fee for the production of carbon dioxide and other greenhouse gases. Without burning gasoline or diesel, the undesirable environmental effects of our farms could be significantly reduced.

Corporations that used to manufacture farming chemicals like fertilizer, pesticides, and herbicides on industrial scales have largely become biotech companies working to create plants that tolerate droughts, produce better vegetables, and don't get knocked over during a storm. I am talking about the seed companies like BASF, Bayer, Dow, DuPont, Syngenta, and Monsanto. There are also a few smaller, but still sizable, agriculture companies in the business as well: Arysta Life Science, Sumitomo Chemical, Nufarm, and Makhteshim-Agan.

Recently, I mingled in the crowd at a political rally in New York City. The theme was anti-GMO—that is, everyone there was against including genetically modified organisms in our food system. As you may know, until recently I was reluctant to embrace GMOs (specifically genetically modified foods, or GMFs) because I was not confident that we could know the impacts of these novel organisms on our ecosystems, agricultural or otherwise. After visiting both Monsanto in St. Louis, Missouri, and the Boyce Thompson Institute for Plant Research at Cornell University in Ithaca, New York, I have changed my mind.

Nowadays, researchers can assay genes at astonishing speeds. At the labs I visited, scientists and technicians can get the complete gene sequence and many of the novel molecules extant in plant cells literally ten million ($10^7$) times faster than they could just six years ago. This technology has enabled agricultural biologists to assess precisely what each gene in the plant's genome does. The research is so diligent, and the checks and balances in the U.S. food system so well established, that I'm not concerned about GMFs the way I once was. And the upsides are significant. Modern novel crop varieties could offer a faster way to make drought-resistant or salt-tolerant plants, to get to bigger and more efficient yields; to have a whole portfolio of new varieties ready to go as the climate changes in large agricultural production areas. This kind of work is already under way, and I think it's going to be an important tool among the many needed to improve and expand our food supply in a changing world.

A lot of the resistance I see to GMOs has less to do with the perceived safety of the food or the ecosystems than with a basic mistrust in large corporations—especially large industrial chemical corporations. Monsanto's high-profile lawsuits, combined with that company's manufacture of both the creepy herbicide Agent Orange many years ago, and the currently manufactured herbicide glyphosate

(Round-Up), have stirred suspicion and hostility. People romanticize the idea of the small farmer, and inherently doubt the values of the giant, faceless company. But the business world is changing, too. Corporations everywhere are merging and becoming more consolidated. Just compare how many car manufacturers there were eighty years ago compared with today; once there were dozens and now there are just a handful. The same has happened in farming. For many people, it is perhaps a regrettable change, but it tells us nothing about whether or not GMFs are a safe and useful technology.

At any rate, after being at the anti-GMO rally for a few minutes, I can tell you that it's not the big corporations that scare me. It's these people at that rally. When one speaker insisted that the U.S. president Barack Obama was part of a conspiracy sponsored by large agriculture companies to control minds—and received a great many cheers—somehow the passionate man at the microphone crossed a line for me. To me, conspiracy theories are lazy. It's a way to blame one's problems and the problems of the world on someone else, some mystery group that's out to get ya'. I am pretty sure that most of the problems of the world just happen from all of us trying to make a living here. I'm all for raising legitimate questions, but these people seemed to be woefully uninformed and obsessed with finding a scapegoat for what they perceive as society's ills. I'm pretty sure that working together we can address climate change and transform the world. I'm absolutely certain we cannot succeed by turning our backs on technology.

What these large agricultural companies are actually doing is seeking ways to do more with less, on every farm in the world. Making a profit is a part of the deal, of course, but that is true for every business. It is true even for the organic farmers (oh, my). The modern researchers are working to breed or transgenically modify plants so that they are resistant to disease, armored against pest insects, tolerant of low rainfall conditions, can be raised without harrowing and tilling soil, can grow in the presence of herbicides, and can, in some cases, grow in water so salty that few other plants can grow there at all, let alone useful crop plants.

The scientists I spoke with at Monsanto and at Cornell University believe that the total area of land under agricultural production will actually decrease in the coming decades, forced by circumstance and aided by a number of advances. GMFs are not the only important advance, but they will play a role. The numbers those scientists quoted consistently were these: Right now, we farm on about 11 percent of the dry land on Earth. By 2050, we will be farming on just 9 percent. That's a decrease of 2 percent, which translates to 45 million hectares (111 million acres) to feed 9 billion people. The decrease means that farmers can give up some land—yielding it to urban growth, to new forest tracts (afforestation), or in some cases pulling inland away from areas that climate change has rendered unsuitable—and still produce enough food for everyone. The scientists and farmers working are confident that this reduction can be achieved in a sustainable fashion by means of "Sustainable Agricultural Intensification." They believe that they can improve farming by introducing new breeds and varieties of crops that produce more food on less land.

Reducing the footprint of agriculture could also free up land to be reforested or reestablished as riparian or coastal wetlands, which could allow an increase in biodiversity in the coming decades. It's an outcome that people at that anti-GMO rally probably would not expect. Inevitably, there will continually be new pests, new plant diseases, and novel varieties of noxious weeds, but the food researchers think they can stay ahead of these emergent pests, pathogens, and weeds. I now think they're right.

Technology is important for feeding humanity on another fundamental level as well. In the developing world, a great many problems could be solved if we in the developed world can find ways to improve infrastructure, pave roads, and electrify their countryside. With improved roads, goods can be brought to market more easily. With reliable electricity, farmers have access to good refrigeration, well-lit warehouses, and storage facilities. Furthermore, when farmers around the world are connected to the Internet, they can coordinate their plantings and take best advantage of the markets.

About ten years ago, I had the chance to interview several people who promoted sustainable agriculture. To support free trade of coffee, I made a contribution and received a radio manufactured for farmers in Africa. It's not like most radios you may know. It's powered either by a small solar panel or by you. There's a large crank on the back that a farmer (or his enthusiastic kid) can turn, storing energy in a spring. With a durable generator inside, the farmer can hear radio broadcasts of the latest coffee prices. This enables him or her to know the current value of her or his crop. It's much more difficult for a charlatan to convince a farmer of a false market price and make a profit by selling at a higher price because the current market price was theretofore kept secret from the farmer. It's using information technology to make coffee farming more efficient, smarter, more profitable for those who need the profit most.

Just think if that information seed were planted everywhere, and every farmer had access to the best and latest information on weather, farming techniques, and water availability, as well as the latest or current market prices. It would improve efficiency and feed more of us while consuming less of the world's resources.

While we are looking for ways to improve agricultural efficiency, I want to point out it is probably not a good idea to provide economic incentives for enormous tracts of land planted in only one crop. Such large-scale monoculture probably contributes to the current difficulties we're having with honeybee colonies and monarch butterfly populations. Today's commercial bee colonies are itinerant workers, trucked from one uniform set of crops to another. They are expected to search for nectar in fields of a single crop after long journeys on the back of a truck. Furthermore, honeybee populations have become inbred and less genetically diverse. They're far more vulnerable to disease and parasites than they used to be. No wonder the colonies are in rough shape. Monarch butterflies can't find enough to eat if their milkflower plants (milkweed) are wiped out over a large area by effective popular herbicides. Our modern farming practices have to become more sustainable. As a guy who is not a professional farmer, I can only observe that it looks like we need to merge some traditional agricultural practices, such as mixed use of farm and pastureland, with modern biotechnologically produced crop varieties.

We have to do more with less on the farm—and we can. The lives of people in the developing world are improving daily. They are living longer. They are getting more nutritious food. We eat twice as much cereal grain as we did fifty years ago. The Green Revolution in India enabled that country to feed a billion people on an area of land smaller than what the industrialized U.S. uses to feed a sixth as many. But many aspects of today's food supply are not sustainable. Worldwide we are consuming three times as much meat as we were a half-century ago. We are catching six times as much fish. We cannot continue to farm in ways that spew greenhouse gases. We cannot take from the sea without allowing its fish and shellfish populations to maintain sustainable levels.

Many of the ideas I've discussed in the earlier chapters merge here: carbon-free vehicles, efficient production of clean water, clean energy generation and use, effective energy storage. All of that needs to come together with smarter ways to utilize the land and targeted ways to improve our crops. Let's enable innovation, and provide good roads and good refrigeration. Let's waste not so that we want not. I think farmers and eaters of all persuasions are down with this idea.

# 26

# BRINGING IT ALL BACK HOME TO BILL'S HOUSE

One of the hardest things about discussing energy and climate change is trying to grasp the scale of the issues. People talk about climate change in global terms, because it's a global problem. Even when they talk about specific ideas like wind power, battery storage, and a better electric grid, they tend to describe them in sweeping, nationwide or industry-wide or planet-wide terms. The conversation keeps turning to billions of people, millions of square kilometers, trillions of watts, billionths of a meter. I'm not pointing fingers. I do it, too, and as you probably noticed I did it plenty in the earlier chapters of this here book. I think it's unavoidable when the issues are so big.

But as I mentioned early on, the issues are also very small. They are about technological changes that will affect all of our lives, and about lifestyle changes that add up to collectively change the whole world. I know, because I've been experimenting with and experiencing a whole lot of them. I am working hard to do more with less in my very own house as a means to understand existing technological opportunities in communities all over the world. So I'd like to step away from the enormous scope of changing the world, for a while at least, to my house, my home. I call it Nye Labs. It's like Earth, only somewhat smaller.

My house is my laboratory. I'm not talking about a single room in my house; I'm looking at the whole thing with its many subsystems. I have solar electricity, solar hot water, active electronic water-use reduction devices, special windows, specialized insulation above and below the main living space, passive ventilation, active and passive irrigation of my xeriscape (dry-scape) front yard area and of my food-producing garden, and of course, good old (new) fancy lightbulbs. All of these are existing technologies. Generally, each involves a bit of design and a bit of good materials science. If every building we have had some of these features, we would have well over 10 percent of the total amount of energy we use returned to us for free. That would represent billions of watt-hours available to us all (some big numbers again). Here again, the longest journey starts with a single step.

The house I own is in Studio City, California. It is, as certain denizens of the San Fernando Valley say, like totally in the Valley (totally). For a number of reasons, which I'll describe here, I set about taking advantage of a great many currently existing technologies to save energy. I'm the first to admit that I am a tinkerer. I have gotten caught up—some might say obsessed—with the idea of doing more with less. For me, this came to mean conserving energy wherever possible in my house. I am continually designing gizmos to conserve energy and produce energy from the Sun for electricity, space heating, and hot water.

I understand that I have become a minor celebrity; I was able to earn enough money to invest in these ideas. I have no kids in college to help spend my savings. But here's my claim. With these investments, which amount to less than the cost of one of the small swimming pools that you see in my neighborhood, or less than one of the very nice sport utility vehicles that are ubiquitous on my street, I have systems that save a lot of energy and a great deal of money. Unlike a nice car, the investments pay for themselves over a decade. It's largely a matter of seeing the opportunities and taking the time. And for me, let's face it, it's just big fun.

For example, it's just fun to get an electric bill for ten dollars every sixty days. It's fun to drive an electric car and spend about a fifth for your electricity than you would have spent on gasoline. It's fun to watch the electric meter go backward during the day. It's lovely to watch the indicator light on the water heater controller turn off during your shower because the over-hot solar-heated water has made it from the storage tank to your shampoo. It's just a little frustrating to reach for the light switch and realize that it's not that you left the lights on, it's that there is so much free light coming down from the roof through elegant dome-shaped lenses. These are good problems to have. While design and planning can be a lot of work, it's also engaging and fun.

My beloved neighbor Frema, who was eighty-six when she died, told me that when she and her husband first moved to this part of the San Fernando Valley it was quite rural. There were orange groves everywhere. Today it's part of the horizon-to-horizon development that is Southern California. Most of the houses on my street were originally built in the 1950s, long before anyone in sunny SoCal thought about insulation, saving energy, or a long-term unrelenting drought. Oh, but we all do now! I know a few of my neighbors pay water bills of over one thousand dollars a month. You can go through quite a bit of cash living like that.

I am a product of the era of very bad pollution in the United States, when air killed people in their sleep, just as it did in the town of Donora, Pennsylvania, and rivers were so polluted they caught on fire, like the Cuyahoga in Ohio. I attended the first few Earth Day events in Washington, D.C. Since I was in engineering school in the 1970s, there was an emphasis, albeit a small one, on saving energy, on doing more with less. My *Principles of Heat Transfer* textbook, which I still treasure, has an extensive section on solar energy for domestic space and water heating, for example. And this was forty years ago. I often shake my head at how little we've taken advantage of these engineering principles. I met Frank Kreith, the textbook's author, at an American Society of Mechanical Engineers (ASME) meeting, when I was a student. It was a very brief encounter on a very busy convention floor. I tried to explain how appreciative I was of this section in his book. He grinned and said, "Thanks—now go do it." That stuck with me a bit. I got around to using his ideas some thirty-one years later.

# 27

# ¿QUIÉN ES MÁS VERDE—OR, KEEPING UP WITH THE BEGLEYS

For professional television reasons, I moved to Los Angeles in 2001. After living in a very nice apartment for a few years, I decided to invest in a house. Now, I admit I was influenced somewhat by the drive deep within each of us to keep up with the Joneses. But in this case, as you may know, the Joneses are the Begleys, Ed and Rachelle. Ed Begley Jr. is an actor renowned for his energy and consumption-conscious lifestyle. When I moved into my house, I noticed he had solar panels on the roof of his house: He had 9 kilowatts worth. And that was just the start of things. There was a hot water system and cornstalks as high as an NBA center's eye.

Right away, I admired many of his choices. I wanted to keep up. It led to a friendly competition: Who has a greener home, Begley or Nye? I became a regular on their show *Living with Ed*, about how Rachelle, who is generally very supportive, has to endure the odd lifestyle choices of a rabid, some might say over-the-top, conservationist like Ed. This fall, the Begleys will manage to move about a kilometer away to an even greener, more conservationist home that they had rebuilt from salvaged materials and recycled everything. I am still competing with him, though. Do I commit to a gray water system? Do I excavate my whole backyard to install a cistern? *Hmmm* . . . Most of the time, I affectionately-cum-derisively call Ed just "Begley," as in, "I'm watching you, Begley . . . I'm watching you!" We have fun with it all—generally.

Let's start out back. Instead of actually having a usable two-car garage, Ed and Rachelle could fit only one car in that structure, because in the place of a second vehicle was an orderly arrangement of very large and heavy batteries. These things were massive. You could move them only with a forklift. They were resting on conventional forklift pallets and strung together with cables as big as your wrist. His wife Rachelle was relegated to having to park her car on the street, because Ed's electric car was in the garage. (As of this writing, they're still married.)

While I was taking care of unpacking and storage, I took stock of all Begley's installations, and realized that I had the means to pursue some short, medium, and long-term investments of my own that would make my house as or more efficient than Begley's.

### Natural Convection Out Back

So many garages get filled with stuff we hardly use, items we could get rid of, but don't quite. Out of sight; out of mind. But by making the garage area inviting, I came upon another energy-saving idea. My garage can get very warm. By that I mean unpleasantly warm—just stupidly hot, actually. You don't want to be in there; you don't want your car or bicycle to be in there. I considered cutting a major hole in the wall and installing a window-style air conditioner. After the pain of hacking into this otherwise perfectly good structure, that machine would demand a lot of electricity. So, I didn't want to go that route.

Because of the prevailing breezes, any ventilation system I put in would probably have to induce flow from south to north, so I just went with gravity. I cut a rectangular ventilation hole in the wall, between the studs, low on the south side, and another hole up high, near the roof on the north side, high enough to avoid lawn mower clipping output. I put in attractive-enough grated vents designed for attic ventilation and trimmed them with varnished wood.

The south side is always, always warmer than the north, because of the Northern Hemisphere position of the Sun. For those who might not trust me on this, in, for example, the movie *High Noon* with Gary Cooper, the Sun may be as high as it's going to get on that day of reckoning, but in the wild, wild west, the Sun is always in the south, because we're north of the tropics. I've met a great many people who quite reasonably, but quite incorrectly, presume the Sun is straight overhead at noon. Not so; it's astronomy; it's science. Speaking of which, I hung two inexpensive dial thermometers next to each vent, both for visual interest, and to prove to myself that I wasn't crazy (at least not about this).

Sure enough, the roof gets very warm. The air high in the garage is forced out of the upper vent by gravity. There is a flow of air from low to high every day. The thermometers show a difference of about 1.5°C (almost 3°F) almost all the time every day. This difference puts gravity to work for me. It's convection—natural convection. Now, you've probably exclaimed, "Hot air rises," any number of times. It's true enough on Earth, but not because there's something about hot air alone that makes it go up. The warm air is driven up by gravity. Without gravity, we would not have natural convection. It's the cool air low to the ground that squeezes the warm air up and out of the garage. There's a flow all the time, even in the gentle Southern California winter. This is an engineering solution to a small but nevertheless noisome problem. I did not install an air conditioner, and the garage space is always cool enough to work in. It was my first energy-saving maneuver. By the way, I like us all to distinguish between natural convection caused by warmth and gravity, and forced convection, cooling induced by a fan or by blowing gently over a hot spoonful of soup. One form of convection is free. The other costs you calories in lung power.

### Measure Twice—It'll Fit Nice

As you might imagine, a guy like me spends a lot of time in his garage. I am forever tinkering, and maintaining, and thinking deep thoughts about the baseball score on the radio. To make room for my vehicles, the car and, *ahem*, four different bicycles (I ride only one at a time; in that sense, bikes are like shoes), I have made use of the space between the ceiling and my head, and between one wall and another. To get everything to fit, I built my own overhead shelves and my own workbench that is just the right length to fit between a tool chest, drill press, and table saw. I varnished it, and put finish washers under all the screw heads. And you know why? Because it looks better.

My engineering colleagues and my beloved Begley often don't seem to care too much about how things look. *Arrrghh* . . . why not make them nice? *Sigh* . . . . And besides, a commercially available workbench in a standard size would have either been too long or frustratingly short. It's the kind of thing that saves energy because you do not have to go elsewhere to get things done. You do not put off maintenance jobs because you have no room to work. Now I admit, this may be only a tinkerer's or engineer's perspective, but efficient workspaces, like an efficiently laid out kitchen, save energy.

### Don't Hesitate—Insulate

If you live in a house, insulate your attic or space below your roof. Although it is indeed sunny in sunny Southern California, it's what climatologists call "semiarid," half dry. The Valley is not far from an area we call the "high desert," which is cloudless and therefore well exposed to the icy cold blackness of space. In the winter, it gets noticeably cold at night. On just a few mornings each year there is frost everywhere and ice in the gutters by the curb. Then again, in the summer, it is often over 39°C (103°F). Without insulation, my heating bill, and especially my cooling bill, would be a great deal higher than they are today. I blew in insulation with a fan-driven funnel machine specially shaped for this purpose. That's easy. The technology is common, and any service people you might need to install it are skilled and reasonably priced.

But in order to keep up with Begley, I took another step, which was unusual at the time and based at least in part on technology derived from the space program. Skilled guys spray-painted the underside, the inside that is, of my roof with a silvery shiny paint that includes micro-beads of a ceramic material. It reflects heat from the underside of the roof. I cannot help but note that there are many claims and Internet reports about this paint or this material. It helps the heat balance of my house by means of radiation rather than by conduction. It's paint. It's very thin. It provides hardly any insulation in the way a blanket does. Instead, it reflects radiant heat energy.

Heat radiation is just light at a wavelength longer than our eyes can see. (By the way, owls and a few others out there in the dark can see in the infrared.) The coating is applied under the lumber that makes up my roof. It's above the insulation above the ceiling of the main floor below. Any heat that makes it through the insulation would radiate to the roof slats above, where it would conduct its way through that lumber, through the shingles and out into the atmosphere and sky. With the shiny paint, some of the infrared energy is bounced back down, keeping the insulation and house below just a tad warmer than they would otherwise be.

The shiny paint works the other way, too. When the sunlight is beating down on my roof in the summer, it gets the lumber of the roof warm. It would radiate that heat eventually to the insulation and the ceiling. With a shiny undersurface, less heat is radiated down onto the insulation and ultimately the living spaces below. Not only do very dark or black surfaces absorb more heat, they also radiate more heat. In other (more) words, not only does a black car get hotter in the summer; it cools off faster in the winter than does a light-colored car. You can feel it. On a hot day, put your palm on the roof of a black car, and then on the roof of a white car. Careful not to look suspicious or furtive. If you're out and about on a cold day, try the same thing. You'll also notice that frost forms on the roof of a dark-colored car sooner than a light-colored car on a cool night.

The confusion for many people about the radiation-barrier paint comes when they try to evaluate claims that the paint itself is an insulator. If that were true, by logical extension, you could just paint interior walls with this stuff and feel as toasty as being wrapped in a comforter or quilt. It doesn't quite work that way. The shiny paint affects the radiation rather than the conduction of your structure's heat. Under the roof slats it's quite effective. Under a layer of designer paint on a living room wall, not so much.

As we like to point out in science, there is no such thing as cold, just the absence of heat. The same property that makes black surfaces absorb or radiate heat effectively is the property that makes it appear black in the first place. As a first cut at understanding, it's that simple. By the way, if you get to thinking about it, chrome is an illogical coating for a car tailpipe. It's going to get hot and stay hot. That's generally why so many chrome tailpipes are discolored: The plating can't stand the heat. But looks matter a great deal in car sales, especially on the showroom floor.

### Recklessly Enabling Aboveground, $CO_2$-Emitting Fuel

My house was built in 1951, when keeping warm in the winter meant getting closer to the fire. I still have a fireplace. It's been there for more than sixty years. When I lived in Seattle, we had an earthquake in 2002. It damaged a lot of brick buildings and broke my chimney loose from the house. So I wanted to make sure my Los Angeles chimney was earthquake ready. And, oh yes, I thought quite a bit about just taking the whole thing down and having a blank wall in its place. But as a pretty good Boy Scout and camper, I very much appreciate an occasional fire. So I had the chimney refurbished. A skilled crew came and repaired or tuck-pointed the brick with special fireproof fiberglass-reinforced mortar. They smoothed the interior walls of the flue as well. After a brief sales pitch, the boss sold me a reflector. It's a piece of stainless steel that rests on a slotted stand, like a European toast tray. That's all it is, a sheet of steel on a stand. I tried it, and I was amazed at how much more sensible heat came into the room.

The reflector was only 60 x 60 centimeters (2 x 2 feet), but it really increased the heat output. I didn't quite believe it, so I set up an infra-red ear thermometer on a lab stand about knee high in front of the fire to get a sense of the temperature change with and without the reflector. Without the reflector, it got to about 36°C (97F). With the reflector, it's about 42°C (108°F). That is the difference between running my central heating system and just leaving it off all night. The first reflector didn't look all that great; it looked like an afterthought . . . that is, instead of a brilliant engineering solution to a subtle problem and a way to revolutionize all of humankind's space-heating procedures (?!).

I pressed a large piece of cardboard into the back of the fireplace (when there was no fire burning) and produced a template of the exact shape of the back wall. The custom-fitted reflector I had made now goes edge-to-edge, bottom-to-top, and corner-to-corner. It really puts out heat now. I can tell you from experience in the avionics industry that when you can hold your hand on a piece of metal for only a mo-ment . . . like when you place your palm on the autopilot electronics black box, for example . . . and you can only keep it there for less than a second before your reflexes make you pull your hand away—that surface is about 60°C (140°F). Well, the back of my fireplace gets well over that, and the reflected heat is just wonderful.

I made another important, straightforward, but somewhat rare modification to the fireplace that every homeowner in the world should implement. The air that feeds the fire now comes from a duct connected to the outside of the house, from behind the chimney. The chimney guys made this little modification to the so-called "ash dump" chute. The fire does not draw air from the room; its oxygen comes from outside of the house. It makes all the difference in the sensible heat you'll feel. If you allow the fire to draw air from the room, you are convecting warm air, and the heat that you paid for, right up the chimney. Oh my. An interesting feature of my house also is that it's so well sealed up now that if you run the stove exhaust fan—which is several meters from the fireplace—while you have a fire going, smoke will get pulled into the room. The house is that well buttoned up and I guess the stove exhaust fan is that strong. The smell of a fire is often comforting, but all things in moderation.

Burning a wood fire is old-fashioned, and I'll admit that I bought a half cord of firewood soon after I moved in. Well, it's my policy now to never do that again. Before you bite my head off, I understand completely that the fireplace may not be long for my world. Even if I choose only to build a fire when it's raining, it still puts out quite a bit of particulate pollution into the already troubled Los Angeles air. Chemically, I'm just putting the carbon dioxide back into the air that the tree took out of the air to make itself in the first place. There is, at this level, no net addition of carbon dioxide. But the smoke and particulates are undesirable, and worthy of strict regulation. It's an indulgence for me, and an infrequent one.

Before I let go of this, I'll also mention that for the last nine years, I've burned only wood that fell locally. By locally, I mean on my own street, my block. So far it's turned out that I have enough neighbors trimming and felling enough trees with enough sizeable branches that in nine years I haven't purchased any firewood. So when someone is chainsawing away somewhere on my block, I head over with a wheelbarrow and bring home a few large logs. Of course I limit myself to very few fires a year; I don't burn that much fuel anyway. The best time for a fire is when it's raining. Since I was a Scout I've enjoyed chopping and splitting firewood. Maybe it brings out the Abe Lincoln in us (me). It's quite satisfying compared with, say, sitting all day and typing a book. In this meantime, I'm keeping my options open, but I have a feeling I'll quit the smoke-making habit pretty soon. I can tell stories of the great pleasure a fire brought and how I adversely affected air quality for my neighbors. But at least that good firewood didn't go right to a landfill where bacteria would have broken it down into carbon dioxide with no sensible heat for us people. *Sigh* . . .

A future kid might retort, "Well, Uncle Bill, why didn't you build a plant to recover the methane from the bacteria breaking down your dead wood and use that for the fabrication of graphine desalination screens?" Right now, I don't have an especially good answer. Addressing climate change is a process.

## Efficiency That's Airtight

As I just mentioned, these days my house is very well sealed. But a few years ago, I had a great many small heat-and-cool-robbing air passages. When it came to leaks, once again it was good ole Begley who hooked me up with a company that specializes in house insulation and looking for those costly leaks that let warm out when you're trying to keep warm, and warm air in when you're trying to keep cool. I recommend this, if you haven't done it already. A company set up a fan in my front doorway and sealed all around it with specially shaped panels and adhesive tape. Then they moved around the house squirting a very fine powder in areas that looked like subtle trouble. It turns out that a great many duplex wall outlets (places where you plug appliances in) can communicate leak-wise with the outside. It's an easy leak to stop.

We found a few of these and caulked them up. It made a slight difference in my energy use and energy bills, I'm sure. The big idea this company had was to reduce the heat lost through the floor, I mean through the bottom of the house. In my neighborhood, none of the houses have basements. They were cheaper to build that way, when the Valley was just starting to sprawl. We also live in an earthquake-prone area. And in several very narrow tracts there are underground streams. When it rains on the nearby steep hillsides—not so often, these days—well, it's not the best thing for a basement.

If you're motivated, you can crawl under your house and install alarm wiring, Internet cables, or control wires for various instruments just above you. I've suited up in coveralls and a tight-fitting watch cap, been there and done that. When these houses were built, vents were put in just above ground level to allow air to circulate under the floorboards. And circulate it does, or would. After doing some calculations on how much air circulation a house like mine really needs between its foundation and the house itself, we sealed up about three-quarters of the vents and covered the bare soil with a layer of white plastic sheeting. The white does the same thing as the silver in the attic. It reflects radiated heat. Furthermore, it makes working down there so much easier. You can slide around like a kid on a skating pond. Everything stays much cleaner, which I attribute to less wind, and less dust stirred up, when a worker like myself is down there crawling around.

Oh, and while we were at it, the crew brought in some experts who are skilled at removing asbestos. Left over from some different heating scheme was a solid asbestos pipe, about 2 meters long and 20 centimeters in diameter. I have to say, it was pretty. It looked like ivory or alabaster. But handling it or especially sawing it and creating dust, is quite dangerous. The crew sealed it in heavy plastic sheeting and had it hauled to a special depository in the Arizona desert. Although it is literally solid rock, it's naturally fibrous and had a great many industrial uses for years. Asbestos is mined from Earth's crust, so burying it is not just an out-of-sight-out-of-mind practice, it's quite logical.

With all this, I had spent a relatively small amount of money, less than a family vacation to a nearby beach hotel, let's say, and I had improved the energy conservation characteristics of the house. The insulation, the radiation barriers, and the caulking were all good fun. The big expenditures were yet ahead, each cost being closer to buying a modest new car.

### How About Some New Windows?

The most significant change I made, yielding the biggest improvement in energy conservation, was to put in new high-performance windows. To many, windows sound or appear pretty low-tech. Windows? Really? Why bother? A window is a huge hole in the wall through which a building either gains heat from the Sun or loses heat to the icy blackness of space. The more we can do to control heat gain and loss the better.

If you've never tried this, try it tonight. Open the curtains or blinds so that you see the night through a window. Sit still in a chair and turn the palm of your hand to the window. Concentrate on the temperature you feel on your palm. Now turn your hand so that your palm faces the inside, into the room you're sitting in. Now, back to the window, and then back to the room, etc. You'll sense the difference. When your hand is facing the night, heat energy, waves of infrared light, are radiating right off your hand and into space. When your hand faces the room, some of that heat energy is reflected off the walls back to your hand. It's wild.

I paid the price of a midsized car to have a skilled crew measure the window frames and hardware, choose the appropriate configuration of frame, glass, and opening mechanism, then replace every window in the house. Every window! Begley had replaced only a few of his, and only in certain rooms. Ha! Take that Begley! Your heat is pouring in and out through those old single panes! *Bah, ha, ha, ha, haaaaa* . . . .

Another hi-tech trick used on my modern windows is that their glass is rendered to have low emissivity (low-E). They do not emit heat the way regular, untreated glass does. They are coated with a very thin spray of metal dots. The metal dots are too fine and too finely distributed to distinguish with the unaided eye. It's reminiscent of the tinting in a pair of sunglasses. The material lets visible light pass through into the room while reflecting infrared light before it can either escape the room in winter or get into the room in summer. Then, since my windows were custom measured and carefully sealed during installation, they make the house nearly airtight, albeit only when all the windows and doors are fully closed.

The garage, by the way, is a separate freestanding building. It's from the old days, when house lots and garages were fashioned after horse barns. The horseless carriage was kept in its own stable just as the horses once were. The air in the house is circulated and replenished not by accidental leakage exchange with the outside, but by controlled filtered exchange driven by the fan in the heat pump and air-conditioning system.

# 28

# BILL AND ED IN A FIGHT FOR THE SUN

On the north side of my house is a fence providing privacy between my neighbors and me. My side is painted white. I could not help but notice that the flowers, such as the birds of paradise and bougainvilleas, were thriving. It took me a moment or two to realize that light bounces off the fence and energizes the flowers. This led me to think seriously about installing solar panels. I did some research with colleagues from engineering school. I shopped around Southern California and picked a company. They not only installed the panels, they were savvy in what was required to apply for the discount and incentive program that California provides. The state provides these incentives not because they're a bunch of delusional hippies, but because a solar power plant on my roof means that the Department of Water and Power can use that power to provide service to my house and that of my neighbors—especially during the summer, when the systems on houses like mine produce quite a bit more power than they use.

If you want to install solar panels, and I hope you do, the main thing you have to do is get started. I have met many people over the years whose first, and it seems only, question is, "When's the payback? How long before the payback?" It's less than ten years. Most places it's less than seven. In my neighborhood on my house, it was more like six years. As I write, my system has produced 22,000 kilowatt-hours. I used most of that electricity. So far, I have been reimbursed $9,300 by the city.

If you're not quite paying attention, it looks like I've lost money. I spent about $17,000 to put the system in, but I've only gotten $9,300 back. That's not good. But wait, all my electricity for most of each of the last nine years has been free. I would have spent about $35,000 on electricity. So I've come out way ahead. Whoa . . .

For me, this sort of calculating is a fun game, and for me a solar electricity system is a little like a swimming pool. There are people who feel that a swimming pool is the worst possible thing anyone could put behind a house. First of all, it's costly. There will be horrible construction noise for months. Then it has to be maintained. The pH checked. The chemicals purchased. My, oh my, it's just a big liability. And of course, the more serious and inevitable disaster: All the neighbors' kids will eventually climb the fence one night, get drunk or high, or both, then fall in and drown. And it will be your fault. You will become an accessory to murder. The police officers, who come to arrest you, will have to make their way through the angry mob outside. Most of them will each be carrying both a torch and a pitchfork. By this analysis, it's clear that a pool leads you down a long, humiliating road to a certain death sentence.

For other people, a pool is the most wonderful thing a house could have. You'll be able to cool off after a long day at the office or after a long day on your job installing hi-tech windows and solar panels. You'll be able to have pool parties with all the neighbor kids splashing, having fun, learning to love the water without fear, and staying almost effortlessly out of your hair for hours at a time. A pool is an investment that can be viewed either way. So it is with a solar electric system.

For some people, a solar electric system is an eyesore that requires regular sweeping and vigilance. For others, it looks nice, an appropriate addition to any home. No matter how you feel, it can literally pay the bills. I love my system. Speaking of aesthetics, one of the other expensive but important steps I took was to dress up the ends of the brackets that support the panels above my roof. Currently (that electrical pun again), solar panels are mounted on brackets that stand just above a roof. First, we generally want air to circulate behind the panels. Nominally, the cooler they run, the more efficient they are. Second, we want to be able to access the wiring and the panels themselves should they need to be maintained or replaced.

But, as though I have to tell you, Begley's brackets, cables, and panels did not look very good. By that I mean they were a little bit ugly. The cables hung over the eaves. The brackets were rusty. The panels were attached to the roof in a seemingly haphazard fashion. The so-called "rail-tails" were exposed. I'll admit, so were mine, the ones supporting my solar panels.

I had metal spheres fabricated and mounted them on the ends of the rail-tails. To me, they look great. They're stainless, they match the silvery aluminum rails, and they have a hi-tech feel that I feel fits right in. I bothered to do all this just to make it all look finished and, well, better. Take that, Begley!

## Know Which Way the Sun Shines

There are a few other vital details to keep in mind when it's your time for solar panels. You, or a skilled technician, have to assess how much sunlight you have access to and in which direction(s). When mine were put in, it was an elegant combination of hand drawing and classic tech. The classic instrument for assessing sunlight is the Solar Path-finder. It's a hemispherical dome of glass mounted above a card that features lines and curves generated for a specific location. The curves are derived from exactly the same celestial geometry that enabled our ancestors to build sundials to reckon time, calculate the future position of the Sun in the sky, and navigate the trackless seas. You trace the reflected images of trees and buildings that would shadow your solar panel area at every time of day throughout the whole year. It's a lovely instrument.

Being the eccentric, Begley-crushing enthusiast that I am, I purchased my own Solar Pathfinder just to verify the calculations made by the contractor and to learn how it's done. Oh yeah, that learning thing . . . I can't give it up. Nowadays, instruments fitted with hemispherical domes or lenses that work in exactly the same way as the traditional Solar Pathfinder exploit specialized software that assesses the insolation (the incoming sunlight) electronically, without the need for grease pencil estimating. Although a pencil works well enough most of the time.

At my place, my potential sunlight is blocked by two objects, mainly. First is my neighbors' very tall sycamore tree. It blocks my panels, but it provides shade to my house as well as hers. Shade saves air-conditioning energy during the hot SoCal summers. I'm sure we both come out ahead. The other big sun-blocking object is the second story of my same neighbors' house. Now they travel a lot. So I'm thinking, one time while they're out of town, I'll just cut off that portion of their roof and the walls around it . . . . What? It's easier to ask forgiveness than it is to ask for permission, isn't it? This is the kind of thinking that makes its way into your consciousness as you start to see opportunity after opportunity to modify your house and lifestyle, so that you can save more and more energy.

Rest assured that I have not chainsawed my neighbor's roof down. But I'll admit that the idea sure crossed my mind.

## Plugging Out and Plugging In

Then for any solar electricity system, there are the cables—the wiring. At my place, we had to dig a long trench. After all, I have 4,000 watts that have to get from the garage to the house. Once run up to the wall, my cables are connected to a vital device that enables my energy to connect with the electrical grid's energy. This gizmo is the key to any system—the inverter. It converts the direct current (DC) to the alternating current (AC) that is hooked to every appliance in your house and to my Department of Water and Power. The term "inverter " is not as descriptive as it once was, but comes to us from history, and it's clear enough. Okay, I'll explain.

When generating electricity with a spinning or synchronous machine, we get a sinusoidal output, a sine wave. Half the time the energy is flowing one way, the other half of the time it's flowing the other way. Perhaps it's obvious that we traditionally call one direction positive and the other direction negative. In geometry we refer to the two halves of the output as crests and troughs. By electrically or electronically flipping the negative part, the troughs, of the sine wave upside down, by inverting them, we get an all-positive, albeit jumpy flow, marked by moments of maximum output and moments of zero output. It's a string of hump shapes. So by long tradition, that device is called an inverter. And by the same tradition, a device that goes the other way, making a steady DC electrical flow into a sinusoidally varying one, is also called an inverter. It's an electrical engineer's shorthand. They are essential to any system that's going to be hooked up to a standard, continent-wide, electrical grid.

Begley's more than twenty-year-old inverter machine was as big as a refrigerator. It was taller than he is. Mine is quite a bit newer, and is about the size of two stacked shoeboxes. This size difference is a result of engineers' relentless development of small, more efficient electronics. Oh, it's a wonderful thing. . . . Isn't it, Begley? Ha!

There's another piece of this story on our one block in Los Angeles that really points to the future. Another neighbor on our block, Dan, is also an enthusiast with an understanding family. They have a very nice solar electric system; at least it's nice now. For quite a while, Dan would come over and discuss the trouble he was having with his inverter. Some of his panels face south, and another set of panels faces west. His roof areas are that big and that unobstructed. But for a while, his system was rigged with just one inverter. It turned out that the two sets of panels fought each other. The unlit group acted like a big load or resistor for the electricity coming out of the well-lit group. The answer was, at that time, to rewire the system with two inverters. Rewiring and adding an inverter costs money, but the panels were up and in place. Now Dan has two separate power plants on his roofs. It's a clever and elegant solution. This is a step toward the future.

Right now many solar electric systems are rigged with a separate small micro-inverter for each separate solar panel. This way, shade or snow, or my neighbor's overgrowing weed-tree (whatever species that is), can affect one panel without dragging down the whole system. I can easily imagine an extensive system of very small solar panels, the size of a postage stamp, say, and each one would have its own inverter built in. It would be akin to a modern plasma flat-screen television wherein each pixel receives its own color command or signal.

The three systems are all on my block. Every year, the local elementary schoolkids come by and we each give them a little tour of our energy-saving schemes. The kids may not soak it all up, but they can see the energy future literally in our backyards. When it comes to the solar electric systems, the experiences of Ed, Dan, and me indicate how new this technology is, and how much potential it has. So many contractors, entrepreneurs all, have emerged in the last decade that competition will bring the best to the surface. Installers in the future will not wrestle with inverter and panel-connection designs, and homeowners will know better; we'll all be more savvy. I'm sure that soon, the whole panel-efficiency issue will be handled by a distributed DC to AC inverter system. Varying sunlight will be no problem.

### Our Meters Can Influence Us

To connect the inverter to my electrical service main took some crawling under the house and some drilling through stucco. It was quite an undertaking. But like any remodel, once the work is done, everything looks and feels so much nicer. Not only is there no construction noise anymore, there's almost no sound at all. The inverter hums a little, but the panels produce their power silently, like stalwart soldiers or perhaps more aptly, like the shining of the Sun.

When my system was first installed, I had an electromechanical meter. It featured an aluminum disk about as big as a coffee saucer. It spun backward all day! It was ever so satisfying. Now the city has installed a digital meter, with which they can communicate wirelessly. No one has to come by and read the meter anymore. The meter is more reliable as it has no moving parts. But it's not as cool. . . .

In the near future, I hope each and every household of any kind—house, apartment, or motor home—has an easily seen and understood meter inside, maybe in almost every room, right next to a light switch. Then everyone would easily have an awareness of how much energy the house was using. By merely keeping rough track in your head, you'd know how much of that energy use was because of what you were doing. You would either change your behavior or nag . . . er, uh, suggest changes in behavior to your roommate, spouse, or kid. Knowledge is power, specifically knowledge means power saving.

Getting electricity directly from the Sun has a lot of appeal. It's distributed. Each homeowner can control his or her electrical production destiny, while we're all connected to the same grid and share the same goals of conservation and doing more with less. As I ponder this, I often think about a handsome watch that I wear regularly. It's an "Eco-Drive®." You never wind it. You can't wind it. It has an electronic quartz movement powered by a solar cell behind the face. That cell is about 10% efficient. The ones on my house are perhaps 15% efficient.

Well, what if they were 50% efficient? Or 80% or even 90% efficient, and inexpensive as well? We could change the world in no time. We could have enough electricity for everyone everywhere. Labs are working on that kind of photovoltaic technology right now. Every time I look at my watch, I'm reminded of how well it works, and how well its technology could work in the future. It's an example of where we are today technologically and where we could be tomorrow.

As I mentioned in the beginning of this description of my house, I think of it as a laboratory, a real world experiment. I can imagine a great many people embracing one or more of the technologies that I went wild for. With just slight increases in efficiencies, we will change the world.

# 29

# BILL AND ED GET INTO HOT WATER

The idea or, for some, the dream of living completely off the grid has some kind of super-sexy appeal. Well, not for me. I see great value in being able to buy and sell electrical energy from the community. But when it comes to heat, somehow that's another matter. A house like mine is independent. It heats and cools itself according to its own needs. So for starters, I wanted to see what I could do about my domestic hot water. If it worked out I figured I could share my experience with you all.

If you've ever been outside, ever at any time, ever in your entire life, you may have noticed that sunlight can make things hot, things like sidewalks and the side of your face. (If you haven't, consider notifying the authorities. You may be a space alien.) So why don't we use it to heat our homes' hot water? Well, Begley—that's Ed Begley Jr. to you—and I do. It's easy in principle. It's just plumbing!

When I think about plumbing, I am reminded (of course) of Latin class and my eventual visits to Italy. The ancient Romans had almost unlimited energy resources (because they had slaves). They used that energy to build enormous and very effective waterworks. Just imagine a city without plumbing. I get a little queasy just thinking about the smell. Plumbing has enabled us to have very effective hygiene for millions and millions of people living in proximity all over the world. Running water right where you want it any time you need is genius. I remember an old Western movie in which the hand-lever water pump was in the kitchen. In other words, they drilled the well, then connected the hand pump, and then built a whole house around that one piece of pipe. I remember thinking, "That's brilliant!" And it was, or it is. Hot water plumbing has made the developed world wonderful. No wonder the developing world wants and deserves it.

Having hot water at your fingertips at any moment has taken hygiene and our quality of life up a couple more notches. I presume that most of us reading along here are familiar with a hot shower. It's a wonderful experience, we sing, we think, we come up with new ideas, and we get very clean compared with even our most recent ancestors. Cleanliness leads to not-sickliness, and that enhances a society's productivity. Compare how clean a dish gets when washed in cold water versus hot. I'd say there's no comparison, but there is. Hot water is a more effective solvent, and things just come cleaner.

I've traveled a little bit in China. You may recall my experience with the bicycles and the cars. Solar hot water systems are everywhere there. The so-called "solar collector" is the key component. For whatever reason, in the U.S. we call a device that uses the Sun to heat water a "collector" and a device that uses the Sun to produce electricity directly a photovoltaic "panel." In China, solar collectors are just everywhere. It seems like there is a solar hot water collector on every apartment building, house, railroad station, and grocery store roof. I figured if they're readily available there, it should be simple enough to get one in the U.S. So I set out to purchase a solar hot water system.

In Southern California, we have no shortage of sunshine, so I figured it would be straightforward. Uh, well, it is, but it isn't. At one time, in the 1980s, there was a program to promote solar hot water systems. The plumber I worked with told me that at one point he had fifty crews installing systems—not fifty guys, fifty crews of guys—the idea was so well supported. Along with some other neglectful decisions, like "let's not learn the metric system," and "let's take that solar hot water system off of the White House roof," the Southern California program was curtailed. Those early California systems didn't work very well. But it's easy for me to imagine how they could be modified using modern electronic controllers and modern sealed pumps to work better than they ever did.

Doing a little research, you can find that Australia and New Zealand have several international companies that manufacture and install solar hot water systems of various descriptions. I claim that the opportunities here in the U.S. are huge, if we just got to work developing systems well suited to our climates and plumbing standards. In these other civilized countries they produce hot water tanks that have enough fittings, that is, enough inlets and outlets, to connect the tank to a solar collector on the roof of your house, garage, or tattoo parlor. The tank contains a gas-fired or electric-coil heater. When the Sun is not enough, the gas burner or the electric coil boost the heat going into the system. The water gets to the desired temperature one way or the other. This is a wonderfully logical way to do this job: Get as much insolation (sun heat) as you can. If it's not enough, give the water a jolt of heat—a boost.

Unfortunately, these sorts of tanks are not approved for use in the U.S., at least not yet. It's probably because U.S. plumbers and water-heater manufacturers have not gotten around to lobbying for them. At any rate, a solar-boost system struck me as the way to go. Ed had one, but he had disconnected it because of a leak. A leaking pipe can be serious business. If it's a drip behind a wall, it can nourish "black mold," whose spores are toxic and cause all kinds of allergic reactions to a great many people. And of course, you're wasting water and spending money you didn't know you'd spent until your utility bill arrives by e-mail or post. Also, Begley's system was on the outside of his house. Birds had attacked the soft neoprene insulation, and it looked like it—I mean it looked like it had been under attack. C'mon Ed, this looks awful . . . . Hence wife Rachelle's continual arms akimbotic state (*sic*).

In practice, there are two varieties of solar hot water collectors. There's the simple flat-plate style. It's a zigzag of copper pipe or tubing in a glass-topped black box. (Pipes and tubes are the same, but different. Pipes are referenced by the inside diameter or I.D. Tubing is referenced by its O.D., outside diameter.) The glass is usually a "low-iron" type, which has "low transmittance in the long infrared"—I mean, it conducts less heat and is a better insulator than regular window glass. Next, you contrive a means to pump water through the zigzag, which is sitting on the roof angled toward the hot sun, and the water in the pipe gets hot. You store that warmed water in a tank, and you have preheated warm water ready to be boosted to hot. On many days during the summer, my system goes way past "warm" to "very hot." At the other extreme, such a setup is not so good, when the temperature plummets in a polar vortex, bitterly cold weather event. The pipes are too exposed, and they can too easily freeze.

Because of their efficiency and because of temperature extremes, in most other civilized countries, these sorts of flat-plate black box collectors are outcompeted by the more sophisticated evacuated-tube collectors. By another nomenclatural tradition, "evacuated-tube" collectors are completely different from the "vacuum tubes" in old-style electronics. This style of collector features a black tube of fluid supported in the middle of a larger clear glass tube. The space between the two tubes is evacuated, a vacuum being the least conductive medium you can come up with. The only means of heat transfer is radiation through the vacuum, which is slower than conductance even through something like foam insulation. This is the principal principle behind a Thermos bottle. The lower portion of the outer tube is aluminized to form a curved mirror surface that focuses sunlight on the bottom of the inner tube so that it gets heated all the way around, from upper side and under side. The fluid in the inner tube is often just ethylene glycol mixed with water. Don't fret. It's antifreeze just like you use in your good old-fashioned, fossil-fueled internal combustion automobile.

The inner tube or pipe forms what's modestly called a "heat pipe." It's a basic-sounding name for a pretty sophisticated device, which is, once again, a consequence of having a space program. Such pipes were first created to move heat around on spacecraft headed to the Moon. Here's the idea. Sunlight heats the working fluid in the heat pipe. It's often a conventional refrigerant. It collects at the base of the tube. Because that inner glass tube is also partially evacuated, the refrigerant boils readily and evaporates at a much lower temperature than it would out in the open air. The vapor rises (gets squeezed by gravity) so it reaches the top end of the pipe. That end is contained in a potable water-filled reservoir or tank. The refrigerant fluid gives up its heat to the water in the upper tank, changes from a vapor back into a liquid, and dribbles back down to the bottom of the inclined assembly. Heat is transferred whenever the Sun shines—and with no moving parts, save the readily available, very reliable air conditioner working fluid. Voilà.

Perhaps one of the reasons you seldom see these systems in the U.S. is that if there were to be a leak between the heat pipe fluid—the refrigerant fluid—and the potable, drinkable, showerable water, it would be more than a little bit toxic. Furthermore, some mixtures taste sweet. That's why you've got to be careful if you let antifreeze spill on the garage floor. A dog will lap it up and get very sick. That said, in other countries all over the world, these kinds of systems are used routinely. So it is reasonable that they work fine, if they're made well enough. But the evacuated tube-style heater is more complicated than the flat-plate box, so it will cost somewhat more. Furthermore, in Southern California, it seldom gets cold enough to freeze the whole tank-to-roof-to-tank loop, so the vacuum-in-a-tube complexity may not really be needed.

When it is cold out, and the controller senses the flat-plate box temperature getting too low, my system runs the pump. The circulating water keeps any of the pipes from freezing. That little bit of motion provides the kinetic energy that gets converted to a little bit of heat at every turn, every bend in the pipe, and that keeps liquid water from changing into solid water—ice. There's no issue of potential toxic leakage, because there's nothing but potable water running everywhere.

In the bigger picture, both types of heaters, flat panel and evacuated tube, work the same way: There's a temperature sensor fastened to a pipe next to the outlet of the heater assembly on the roof. It's either right next to the flat-plate black box or right next to the evacuated-tube potable water tank. There's a second temperature sensor in the big primary water tank down on ground level. When the solar heated water is 3°C (5°F) warmer than the top of the tank, a pump runs circulating water from the tank up to the solar heater and back. So as the day goes on, the water in the tank gets warmer and warmer. In the summertime at my place, the water often gets quite a bit above 52°C (125°F), hotter than the standard safe setting. In practice, by the time the water flows from there to a sink it's just barely hotter than 52°C. But it is big fun when all of that heat is free.

If you like to concern yourself with some details, there aren't that many. The pump, for example, is lovely. It uses under 30 watts, less than an old-style reading lightbulb. The spinning impellor is not on a shaft, as such. It's spun in its housing coupled to the motor by a pair of magnets. There is no slipping O-ring to wear out and leak. It just runs and runs.

Most hot water systems, regardless of the kind, have an expansion tank. This metal tank—somewhat smaller than a scuba tank—is fitted with a rubber bladder inside and pressurized from below with air. You fill this small tank with air in the same way you might fill your car tires. This compensates for the water in the hot water tank expanding. It's exactly the same phenomenon that is causing Earth's sea level to rise. The expansion tank keeps the hot water from expanding backward into the city water main, etc. It's another industrial-engineering trick most of us take for granted. It works because air is so compressible, while water is not. The air performs exactly like a big soft spring. The tanks sit there with no maintenance for years on end.

Having hot water is great. Having it arrive at the tap instantly is super great. But all of us have turned on the hot water tap . . . and waited . . . and waited for hot water to get there. The reason it takes so long for hot water to arrive in many or perhaps even most of our houses is twofold. Hot water tanks are often placed in the basement or outside in a separate enclosure. The hot water has to travel from the tank to the faucet valve. But equally important is or are the pipes themselves. The hot water gives up heat to the copper or plastic pipe as it makes its way under pressure from the tank to the sink or showerhead. Even if you are using an inline or tankless hot water heater, there is almost always a considerable length of pipe that has to be warmed, before your shower is comfortable enough to stand under.

Many modern houses have a parallel hot water line that runs from the hot water tank to every sink and shower in the house. I had such a line installed in my 1930s vintage house in Seattle. There's a pump that runs hot water around the loop all the time. I took it one more step and put that pump on a timer, so it only ran shortly before

I got up in the morning and during dinnertime, when one is cooking and doing the dishes. It's okay, but the pump takes energy. The hot water loop, even if it's insulated, is always radiating a little heat into the house somewhere. And often, as a freelancer, I was home when the pump wasn't running. I had to wait for hot water along with the rest of the world. Overall, the system was okay in the winter, not so okay in the summer. Mainly though, for us to do more with less and address climate change, we have to find ways to economically provide instant hot water to everyone. Otherwise, we'll burn through more energy than we should or could, and way worse is that we will all continue to waste water on a grand scale.

Speaking of wasting energy on a grand scale: Most domestic hot water systems around here in the U.S. have a hot water tank that is kept hot all the time. A great deal of the time you are not home, or if you are, you are not in need of hot water. Just imagine looking down on your town with an infrared camera tuned to hot water heaters. You'd see thousands upon thousands of them sitting there hot, and staying hot, and wasting heat. To me, this looks like one of the most solvable more-with-less-problems in the world, perhaps in the universe (wow Bill, pull back).

It is to this end that manufacturers have developed the inline or tankless hot water heaters. They only get hot when water starts flowing, and that only happens when you open a tap or the clothes washer or dishwasher opens one of its valves. These things work very well. But almost always, they are more expensive than the more mature technology of a big ole tank full of water. Inline heaters have some electronics. They have to be reliable (meaning, they have to start and stop every single time). And you have to think a little bit about where to mount them. One of their huge appeals, the instant hot water part aside, is that you never ever run out of hot water. As long as there's a source of heat, be it gas or electric, if the inline heater is big enough to start with, it will always be big enough even for your environmentally politically incorrect hour-long shower.

Now I'm a big-time minor celebrity with no obligations and a deep desire to crush Begley, so I had two separate inline or tankless water heaters installed. The one on the kitchen-sink side of the house is smaller than the one on the shower side, but still having two instead of one about doubled the cost. My idea was to mount each of these units as close to the sink they service as possible, so that the length of "standby water," as its called, would be minimized. Turn on the tap, and in a few moments you'd have your hot water. Well, it works, but only to a point. I still had to wait for what seemed like an inappropriately long time for the hot water to show up. I tried a special temperature-sensing valve that would let water from the hot side very slowly flow into the cold side until the water that was on the hot-side pipes was very hot. Then the valve would close. The idea was that it would maintain a certain minimum hotness on the hot water side. Well, it didn't work, at least not in my system. Instead, it leaked a small flow from hot to cold all the time, day and night. The hot was never quite hot, and the cold was quite warm—all the time.

I found a very good solution to this developed-world problem. The product I settled on is called the Chilipepper. It is an electric pump that is plumbed (connected with pipes) from the hot-side pipe to the cold-side pipe, and it's mounted right under the sink, right next to where you open the tap and get hot water. When you press a very small button, smaller than a doorbell button, the motor turns on and the pump pumps. The button reminds me of the bicycle horn I had as a very young cyclist (little kid). I mounted my hot water control buttons where you don't see them, behind a cabinet door or under a curved section of the cabinets that support the washroom sinks. The control button circuit runs on just 5 volts and 1 milliamp. You cannot get a shock no matter how dripping wet you might be in the washroom.

So here's what happens: The standby water is pumped backward into the cold water line. As soon as water at 36°C shows up, the pump stops. Open the hot water tap, and there it is. Open the cold water tap, and you'll find that the cold water is quite warm for a few seconds. It's a compromise, but a good one. You just don't waste all that standby water the way I did for most of my life, and the way we still do in homes in even the developed world all around the globe. If everyone in, say, drought-stricken California, had a unit or system like this under every sink, we would waste a lot less water. That would leave more water above ground where we can use it to grow crops and as drinking water to stay alive.

I admit I am a hobbyist. I want to see what works and what doesn't, to learn about energy-saving systems and, let's say, write a book like this one. It was a good deal of effort to run pipes in suitable places around my 60-plus-year-old house. I remember one of the plumbers getting his fingers glued together pretty well as he worked in a tight attic space to wrap the pipes with good neoprene insulation. To save time and money, and especially to just get it all done, I crawled under the house to run the wiring for the solar hot water electronic controller. We communicated by tapping and texting. I soldered some wires down there in the dark with just a few centimeters between my forehead and the house floor above. The plumbers crawled in spaces like that for a couple days. It's just hard work. But once it's done, and it works, it's a beautiful thing.

Right now, my tankless heaters are powered by fossil-fuel natural gas. It was the state of the art when I had them installed. If you're going to burn gas, at least do it efficiently. There is still something very troubling to me about using the high-performance energy, electricity, to do nothing but get things hot. Electricity can run computers and phones, after all. Gas-fired systems just make heat and apply it as efficiently as they currently can. I anticipate changing these gas inline heaters someday in the medium term. I played the hand I was dealt when I made this huge modification to my domestic hot water setup.

In nerdy gear-head fashion, I mounted the controller for the tankless heater on a wall, where you can see it as you take a shower. On certain summer days, when you watch that light turn off, you realize that you're living this high-quality life powered by the Sun. It's a good feeling. I put the main controller for the roof-to-tank plumbing in the kitchen and dining area. Everyone can see the little animated diagram and appreciate all the technology there being used to do more with less.

Solar hot water is not going to solve everyone's energy problems in one weekend. But for me, it stands as a classic example of a resource that we've hardly touched. If people went into the solar hot water business, I'm sure there would be quite the market, especially if these entrepreneurs are savvy enough to design heat-boost systems that comply with the crazy quilt of regulations for our domestic hot water heater tanks and ancillary plumbing. Domestic hot water is an area that I believe we could save countless, or at least a huge number of, Joules (or calories or BTUs). This very achievable technology could go a long way to enabling us to do more with less, right here and right there at home.

# 30

# THE TAP IS OFF AND THE GARDEN IS GREEN

While my neighborhood in California (the whole state, really) is in the midst of a serious drought, and we are all conserving water as best as we can, I just can't help but think back on when I had a lawn out back of this house, and how I used to make some money mowing lawns as a kid. It was hot hard work, but the smell of the grass always brought me joy. As a kid or especially as a grown-up, it's hard to beat the feeling of playing ultimate (with a flying disk) or hardball baseball on a lovely grass field. I think there is something deep within us, perhaps dating from our ancestors' time on the African savannah, that makes a grassy open space appeal to us like nothing else. With that said, my house had nice green grass lawns when I moved in. I tore them out—front and back.

Growing and maintaining a grass lawn in Southern California is expensive because of the cost of water, and right now it's downright irresponsible. Where is that water going to come from? What if my neighbors' kids need a drink? Even rich people will soon not be able to buy all the water they want. There just isn't enough to go around.

To keep California's agriculture going, and provide water for sanitation and hydration, we can't afford to pour it on our lawns no matter how much money there might be to throw at the problem.

This was another area where Begley (my rival neighbor Ed Begley Jr.) had filled me with ideas. He grows corn, squash, and tomatoes out back. I figured why not do the same at my place? I could have nice flowers and even some food, if I played my cards right. One productive afternoon Ed drove me to an environmental film festival in his old-by-this-time all-electric Toyota RAV4. The event had a real hippie feel. While moseying around the displays, I met Tara Kola, who is all into what she calls urban farming. She and her team came out and together we designed an arrangement of five brick aboveground planter boxes. I use the term box here to describe enclosures that are solid brick. There's no lumber. (Ed's were rickety stacked paving stones, many of which had fallen out of place from the motion of water and roots. *Ha!*) Tara's company provides a service to people like me. Every few weeks, they come by what I whimsically call Nye Farms and plant seasonable vegetables and fruit.

Between the planter boxes is a brown gravel substance that the folks at the garden center call decomposed granite ("DG"). It's granite with clay minerals mixed in. Over millennia it becomes quite frangible (i.e., fragmented). We packed it down hard. And from to time, I go out and repack it. Today, I find myself owning a tamping tool. It's a flat metal plate on the end of a heavy wooden handle. I thought they were just for professional landscapers, but it's quite handy. You just walk around and pound the ground. Makes the soles of your feet feel like there's an ancient dinosaur walking nearby. The DG is handsome, when all is said and pounded.

As the gear-head, I could not resist rigging up an automated irrigation system. Paul Byrd, a professional landscaper, and I got it at one of the big box home improvement stores. Talk about doing more with less! It has a rain sensor and is programmed for nine separate zones. I can tune the controller to water very specific areas of the "farm." The irrigator outlets themselves are very small, almost tiny, drip outlets. It's not an aboveground, spray-water-all-over-the-place setup. The water just drips, and generally at night, so that not much is lost to evaporation. Right now, kale is a popular leafy green vegetable. Under the growing conditions I've set up, kale is practically a weed, and it grows as big as a small tree. The problem that I work with continually is giving the food away, as I often have a surplus. It can make for good neighbors. I've thought a great deal about just planting less, so that my yield will go down. One season, I let one box go natural. Some wildflowers grew there and it brought a great many happy bees (or they seemed happy to me). But even that takes some paying attention, etc. So far, it's been a good problem to have.

I used to have an all-American lawn out front, the kind of lawn you might see in a Norman Rockwell painting. Keeping it green cost hundreds of dollars a month. I found the right landscape architects. They knew Ed (of course), and they had won some awards for their water-conscious park designs. My front yard is now an expanse of pebbles—that xeriscape that I mentioned earlier. The name comes from the Greek "xeri" meaning "dry." It's a dry landscape on purpose. With an expanse of pebbles to work with, another longtime acquaintance helped out. Linnette Roe is a retired ballerina who is now a certified nursery person. These people babble in Latin, *id est*, they refer to all plants by their genus and species. For example, "I like the feather grass." "You mean the *Stipa tenuissima*? If that's what you mean, just say that."

Paul and Linnette designed an arrangement of succulent plants that is quite handsome. The whole front area is irrigated with the recycled tubing that's manufactured with tiny outlets along its length. It's buried just beneath the soil. It works pretty well. You can tell when an area gets clogged or damaged. Either all the succulents for a few meters around die. Or they go crazy, because they're receiving a whole lot of extra water. It's one more setup that requires maintenance. But when it's working, it works great. I have the pebbles all the way up to the curb. It's just not the best surface to walk on after you get out of a parked car. It's a process. When I come up with something better, I'll change it.

Begley had decomposed granite on the curb. It's okay. But that dust gives it a desert look that doesn't quite go with the street—in my humble, but correct, opinion. Did I mention that it's a process?

When you look at my house from the street, you cannot help but notice a lovely camphor tree out front. In the spring, it smells like Vick's VapoRub. It's a huge money saver for me—the tree, not the camphor ointment. It shades my house in exquisite fashion, and that controls my otherwise outrageous Southern California air-conditioning bill. I have spent quite a bit of time pruning it—the tree, not the bill (well, the bill too, I guess). I admit it's my kind of adventure. I have a very tall "orchard ladder." The idea, if you're into this sort of thing, is to make it open inside like a basket resting on the main trunk. It works. The tree is thriving. But there's some engineering going on below.

I began this effort at the beginning, I mean when I first moved in. The second week I was there it rained and rained for four consecutive downpouring days. Walking around the house in the rain, I could not help but notice that the previous owner had installed some gutters. They were in the back, and they were backward. Instead of the runoff being directed away from the foundation of the house, the downspout put a very large flow right into the lowest corner of the backyard patio. What seemed like a small lake formed there that first week. I thought about water-skiing, but that would mean a boat, and so on (jk—just kidding, as the kids say). The previous owner had put one of those gutters there just for show, to give the appearance that he had addressed the very large runoff that comes to a head in the roof's valleys (a term of art in the roofing business).

A couple of things occurred to me. First of all, I should get that huge puddle away from the house. Water will get in the soil and allow the foundation to shift and slide around, which would slowly pull the house apart from the ground up. Undesirable. But I also realized that I could capture that water and do something with it someday.

I lived in Seattle for many years, where it really does rain a lot. They're not kidding when they joke about it. "I asked a kid, 'Does it really rain here all the time?' He said, 'I don't' know; I'm only eleven.'" A few houses there have rain gutters that appear to be entirely covered with a sheet of metal—and they are. There are a few brands like this around now. Rain runs over the metal cover sheet, and it sticks to it in the same way water sticks to an unwaxed car hood. The lower or outer side of the cover curves down in a shape reminiscent of curled fingers. The water sticks to the metal as it flows over that bend, then it breaks free and falls into the gutter immediately below.

You never ever have to clean the gutters. For me, that's convenient, of course. But it also saves energy. You don't have to spend time cleaning the gutters. You don't have to hire a skilled workman to drive his truck over. You don't have to chase that last bit of debris in the gutter to the downspout with water from a garden hose (water that you have to pay for). The only drawback of systems like this is that they cost more to install than uncovered gutters. There is about one-and-a-half times as much metal per linear meter or foot, and the installation costs about one-and-a-half times as much as conventional arrangements. But they work better. The gutters will be there for many years to come. Once again, it's a product of longer rather than short-term thinking.

When I first got the gutters up, their runoff just ran into the alley and into the street, to wit, to where conventional roof runoff runs.

Rainwater flowed down the street and into a branch of the Los Angeles River, which is an enormous concrete channel with virtually no riparian habitat along my neighborhood's section of the flow. A couple years later, I made a significant modification to this arrangement.

The front downspout is routed to two very large gravel-filled recycled plastic barrels, which are buried beside the tree. You would never know they are there. They're under the pebbles and plants. When it rains, seldom but seriously, that tree gets barrels-full of rainwater. The occasional rain sustains the camphor tree, which in turn shades my house in the most beautiful way. To put this passive irrigation system in was a bit of a project. First, we measured the roof to estimate the volume of water that runs off in a typical rainstorm and over a typical season. Then we rerouted the copper downspout to flow into a plastic irrigation drainpipe. We dug trenches and moved soil around. It's all sized to do the job of keeping that tree alive and thriving. It took planning and engineering. It's completely passive.

Many, many such opportunities exist for all of us, both literally in landscape planning and metaphorically everywhere we use water. I think about it every time I sit on my front porch, take in the scene, and yell at cars to slow down. Ask the neighbors; I routinely yell at drivers. There are a great many kids and dogs around. Sitting outside I wave to my neighbors walking their dogs. I've learned a lot of names that way—I mean, of the dogs. Come to think of it, the humans probably have names, too.

The porch, by the way, is amazing. It's big enough for several people to sit outside sipping coffee and talking about life. Looking back, I was brought up with such an arrangement. The house in which I grew up has a porch big enough for the family to sit together and sip iced tea. When people sit outside on a Sunday morning or any evening, they talk to each other. They establish relationships. My neighbors and I have started a Dutch door trend. With the top half open, you notice things going on. You notice people and their activities. You watch the kids grow up. It brings us together. Neighbors who become acquainted this way split the cost of refurbishing a fence fifty-fifty, for example. Both sides are invested, so both sides have a stake in making it look good. It leads to a higher quality of life for everyone. It inherently leads to cooperation and doing more with less.

I can't say enough good things about the rooms outside of the house, my patio and porch. They save resources, energy, and money. We could emphasize this style of living around the country, and in very short order be doing more with less. Private outdoor spaces are seldom available in dense urban areas. There, we build parks. We have to. They literally give us room to spread out. In any case, I feel small but open space is worth thinking about. How broadly could we apply the science of human needs to future building codes, and can spaces like this be retrofitted to existing houses to save on cooling loads?

I am intrigued by the idea of making large roads narrower, and building thin or narrow structures in the middle of what is now extra-wide streets. We could have zones of narrow houses and storefronts, if we had less traffic or even literally narrower automobiles. Urban density, if properly managed, enables us to create friendlier, safer neighborhoods that use less energy than more spread-out living. It could be one more step in doing more with less.

I meet people who want to own a house in the country. For me, that sort of living is fine for a few days, but soon it feels too isolating. And that business of driving everywhere is exhausting, leastways to me. These differences in desires and lifestyle choices are fine so far as they go. But as voters and taxpayers, we should all keep in mind that if we all really end up paying for what it actually costs to produce carbon dioxide by driving everywhere, living a spread-out lifestyle may lose a lot of its appeal. That idea of having to pay a fee for the $CO_2$ one produces could affect the choices we all make. Suburban neighborhoods could lose their appeal, because it would be just too expensive to get anywhere. We'll see. It's another example of climate change requiring regulation. The less we do to address climate change now, the more regulation we will have in the future.

Back on the porch: Along with the pumped potable water irrigation, I have two 125-liter (33 gallon) rain barrels. Oh yes, I tried using gravity to fill a watering can, and then using that traditional hand-held device to transport water to my plants. I discovered quickly that it takes quite a while to get a barrel to drain. I guess this is why whiskey is made so strong. That way you don't need much of it (*hmmm* . . . ). Now I lower a sump pump into each barrel. I got the pump at my local hardware store. The pump is connected to a regular garden hose. The wire is waterproof (it had better be). I water my vegetables with that water, which is rich in the organic materials that fell onto the roof in the first place. The vegetables thrive. But here's the thing. Right now, I don't have nearly enough rain storage. Begley's new house has an enormous underground cistern. *Arrrghhh* . . . I'm watching you, Begley!

We all can do a lot more with less by setting up what I call nice rooms outside of the house. When I moved in, there was a brick patio. It's nice, but it was usually just too hot to enjoy. I built a pergola. This is the Italian-style slatted covering over an outdoor space. Those slats could be difficult to maintain, by that I mean a pain to paint. But these are recycled plastic. I had to import them from Illinois, of all exotic places. They're made from plastic grocery bags and milk jugs. They're textured to look like lumber painted white. Because of the texture, just about any color of paint will stick to them. They are expected to last three hundred years. I'm planning a party at that milestone.

My neighbor and I used this same textured recycled plastic material to rebuild her picket fence. Termites had destroyed several of the vertical posts. They're now plastic. We just refastened the existing crosspieces and pickets onto the looks-like-wood plastic. Termites don't bother trying to gnaw.

Did I mention that it's a process? I preserved one grassy triangle. I call it the "wabe," after Lewis Carroll. In the middle of the wabe I poured a 20-cm (9-inch) concrete foundation for a sundial. It's a tribute to my dad, who loved them, and my involvement with the Mars exploration program. A few of us got together and managed to enhance the color calibration target for the cameras on the Mars rovers to also serve as sundials. On each of the MarsDials, as they're called, there's a message to the future, "To those who visit here, we wish a safe journey and the joy of discovery." That message means so much to me that I justify the extra irrigation and its cost. I imagine soon I'll have to get over that and reconfigure the wabe—which Carroll defined as the grassy area around a sundial—to still pay tribute to the human spirit, without wasting our precious water.

Speaking of my dad and mom and sunlight, I also have hi-tech skylights. They're domes that incorporate a set of grooves. It's a lens with a cylindrical rather than planar zone of focus. The domed lenses direct sunlight straight down a super-shiny duct to the room below. It still gets me. After almost ten years, I still find myself reaching for the light switch every time I go into that room. The amount of light brought down from the roof is surprising. It's far more efficient than a conventional skylight, because of the lens and because the duct has the mirror finish instead of just flat wall paint, as you might expect below a regular skylight. These things are so bright (how bright are they?) . . . they're so bright that the company makes what they call "dimmers." They're motorized doors mounted in each duct that remind me of an eyelid or a so-called butterfly valve. It turns out that you really do need the dimmers; otherwise moonlight will blast you awake for about a week a month.

These tubes, lenses, and ducts remind me of my parents, because my dad became fascinated with sundials, clocks, and timekeeping when he was a prisoner of war. My mom really encouraged me to do my homework and apply to Cornell University. Thirty years after graduating, I managed to fund and design a clock whose face incorporates a feature that indicates when the Sun is highest in the sky each day. It's what astronomers call "solar noon." Navigators say the Sun "culminates" at solar noon. The Nye Clock on Rhodes Hall at Cornell University indicates solar noon using the same brand of skylight dome and dimmer. I'm also very proud to point out that the controller for this mechanism was designed and built by Cornell engineering students. The professionals at one of the major outdoor clock companies apparently did not have the resources (or understanding) to figure that out. Go Big Red, as we say at Cornell.

Each of the ideas I've presented in these personal chapters—the xeriscape, the irrigation controllers for the reduced-area garden planters, the passive irrigation of a very large tree, the management of shade with recycled materials, the solar hot water system, and especially the solar electric panels—are existing technologies. I did not have to go to a research lab and hire dozens of engineers to produce these energy-saving, quality-of-life-enhancing systems. It all existed. I had the resources to complete each project over the course of about nine years. The big ones—the hi-tech windows, the solar electricity, the solar hot water, and setting up my garage for an electric vehicle—have each paid for themselves already and added value to my house.

I can imagine a program of tax incentives, entrepreneurial loans, and energy-fee structures that would enable a great many of us to try these technologies. They are not boutique solutions. As more people get in the game, I'm sure the cost of each system will come down. I cannot help but feel that if more of us were doing more of this kind of thing, we would have more energy and more resources available to many more of us than we do right now. We would also all, personally, have a more direct understanding of the problems we face and some of the remarkable, ingenious kinds of solutions that are available. We would, at every little turn, be changing the world a little bit, making life a little better for everyone.

# 31

# THE CASE FOR SPACE

Up to now, I've been focused on energy as a very practical, earthly problem—and primarily, that's what it is. But if you know me at all, and I'm guessing you do since you've read this far, you know that among other things I am CEO of The Planetary Society and an avowed believer in space exploration. Ever since I was a kid pushing the plastic launch button in the middle of a big field, I have loved rockets. They're just exciting. After you've seen a few model rockets go, you know exactly what to expect. You set up the igniter wire in the rocket engine. You connect the little alligator clips to the igniter. You step back. You count down; it's a wonderful ceremony. There's a very quick *whoosh*, or *whissssh*. Up it goes. It's fantastic. And for me, it never gets old.

Model rockets may seem a world away from issues of climate change, but I'd argue otherwise. First of all, they are a dramatic demonstration of the ways that we harness energy and bend it to our will: In an instant, you convert a great deal of chemical potential energy to kinetic energy, with a wonderful sound, followed a few moments later by the acrid-yet-sweet smell of gunpowder rocket fuel. Rocketry is related to our global challenges in another way, at least as meaningful. It's all about pushing the limits of human ingenuity. That's why NASA earned a reputation for inventing so many amazing spin-offs. It's not just that rocket science requires solving problems that have applications here on Earth—although that does happen, all the time. In space science, we solve problems that have never been solved before. A country's space program becomes its best brand. Difficult engineering problems inspire the best, the brightest, and often the hard working-est (*sic*). Space exploration is the breeding ground for the Next Great Generation.

Every week I meet someone who asks me something like this: "Why should we spend money exploring space when we have so many problems on Earth?" Typically they are surprised to learn that NASA's budget is about 0.4% of total federal spending. (The Department of Defense accounts for 18%, 45 times as much.) I also like to point out that the dollars we spend in space have proven crucial both to understanding and to solving our Earth-bound problems. Without satellites, we would have no clear picture of global warming and the ways in which human activity contributes to it. We would have no precision farming, no detailed weather forecasts, no GPS for guiding today's shoppers and tomorrow's self-driving cars, trucks, and airplanes. Without deep-space missions we would not know about the runaway greenhouse effect on Venus or the 4 billion years of dry ice age on Mars, both of which have provided deep insights into the workings of our own planet's climate.

All these extraordinary advances rely on our understanding of just what it takes to launch things into orbit and beyond. They require rockets, and rockets require energy, a lot of it. In fact, you can pretty much think of a rocket as just a giant tube of concentrated energy. Let's start with this: The Saturn V rocket that took people to the Moon weighed 2,970 tons at launch. Of that, 486 tons were sent to the Moon. Of that, just 5½ tons of payload (capsule, astronauts, space suits, and some rocks) made it all the way back. In short, the overwhelming majority of a rocket is just fuel.

If you just want to go up and come right back down, the energy required is directly related to how high you want to go combined with how much mass you want to take up there. This is the kind of up-and-down trip that is essentially being proposed by space-tourism companies like Virgin Galactic, which hope to take private adventurers briefly above the atmosphere to where the sky is black and you can see the stars even in the middle of the day. It's the same sort of flight the very first U.S. astronauts flew in the early 1960s. It's the most minimal way to get into space, but it requires serious energy all the same. Getting into orbit takes a great deal more. To send Yuri Gagarin on a single loop around Earth, the Soviet Union had to use a Vostok-K rocket. It was no Saturn V, but still it weighed 285 tons and stood nearly 31 meters (101 feet) tall. It could easily have exploded like a bomb—yet Gagarin flew in orbit on his first flight.

As soon as you light the engine, things get even more complicated because the mass of your rocket changes continuously. The weight of the payload (the passengers and their cameras, let's say) stays the same, but the amount of fuel left inside and the amount of fuel you need to burn to keep accelerating changes with every instant. To solve problems like this, Isaac Newton and Gottfried Leibniz invented calculus. Along with a lot of fuel, you need a lot of math to get into space. The fuel consumption problem leads to the famous rocket equation, which involves logarithms. It all really is rocket science.

If your rocket were perfectly efficient, you would need about 500 million Joules to lift one ton of payload to a height of 100 kilometers (62 miles), which is a common definition of where "space" begins. Keep in mind—that much energy only gets you up; it does not get you into orbit. For that, you need twice as much energy, enough for one round-trip driving back and forth across North America. If you want to go out to a geosynchronous orbit, where each loop around Earth takes exactly one day, you need 50 times the low Earth orbital energy. All of these numbers do not count the mass of the rocket and its fuel. You need fuel to move your fuel, and then you need more fuel to move that fuel. The numbers get enormous in a hurry. That's why rockets are so huge.

That's the energy part of the problem. Then there is the matter of engineering. To deliver energy fast enough to lift payloads and overcome Earth's gravity, we have to use a lot of plumbing. We have the rocket engines themselves, the pumps, the valves, and especially the fuel tanks. All this equipment quickly becomes unneeded as a significant fraction of the fuel is burned, so engineers have developed multipart rockets. They're divided into what we call stages. Most modern rocket designs use more than one stage and often more than one type of engine to deal with the different conditions the ship encounters on its way up through the thinning air and into outer space. On that playground back when I wasn't yet ten years old, my friend Val Brown and I tried launching a multi-stage model rocket. After liftoff, we never saw it again. Getting one stage to finish burning while igniting the engine on the next stage is a tricky, tricky business.

Launching a rocket into space requires detailed information about the nature of the atmosphere at every level. This is another way in which rocketry has contributed enormously to our practical understanding of our planet. The pressure of Earth's atmosphere is greatest at the bottom—that is, right here on the surface. The lowest pressure is in the vacuum of space. In between, the pressure is—uh, in between. The pressure standing still is "static," and it's designated with the letter p. But when you are moving through any fluid, like air, whether aboard a rocket or bicycle, you feel this "dynamic pressure," the pressure of moving molecules. For this we use the next letter of the alphabet, q.

At launch, the rocket is moving relatively slowly through the higher pressure of the lower atmosphere. As the rocket gets up to speed, it is moving through regions of lower and lower atmospheric pressure. At the same time, though, the rocket is going faster and faster. There is a moment during flight when the combination of the (decreasing) static pressure and the (increasing) dynamic pressure reaches a maximum. It is called "max-q," and it is the tense moment when the rocket ship is in danger of being shaken apart. The pressure on the rocket's nose, especially, is at its maximum. Max-q is associated with vibrations, buffeting, and things falling apart-ing. Making things tough enough to survive max-q adds complexity and weight to any rocket design.

The atmosphere is not the only environmental factor you need to consider at launch. You also need to pay attention to where you are on Earth. You've probably noticed that most major launches in the United States take place at Cape Canaveral in Florida. There's a practical reason for that. If you want to orbit Earth above the equator, you want to launch as close as possible to the equator, because our planet's rotation is helping you. The spin of Earth adds to your orbital velocity. In Florida, Earth's surface is rotating at 1,470 kilometers per hour (913 miles per hour) relative to a stationary spot in space. At the equator, the speed is faster still, around 1,670 kilometers per hour (1,040 miles per hour).

So you might ask, why don't we launch from the equator? Well, at the time of the space race the United States was in the middle of the Cold War. Any space capability that this country was going to develop had to be on American soil, but as far south as possible. The U.S. also wanted a launchpad in the continental forty-eight states, accessible by rail car, because rocket parts are big and heavy. That ruled out Hawaii, for example, which by the way had not achieved statehood when NASA was formed. When the European Space Agency was established in 1975, managers did indeed choose a place very close to the equator. In French Guiana, you'll find the Centre Spatial Guyanais, CSG (Guiana Space Center). At just 5.2 degrees north of the equator, it's ideal for taking advantage of Earth's spin. The drawback for ESA is that rocket parts have to travel there by ship, but the location comes with an advantage when it comes to getting up to speed—up to orbital velocity. Sometimes you want to launch a rocket so that it orbits north-south, going over the poles. That requires more fuel for a given payload, because you get little help from the spin of Earth.

These are the basic energy requirements and engineering of a rocket launch. Ah, but after you get in space, quite often you want to come back down, leastways if you're a human or a payload that's worth recovering. To do that, you have to dissipate all the energy that you pumped into the payload getting it up there. That is why reentry is such a fiery process. The easiest way to handle all that energy of motion above the atmosphere is to use friction to convert it to heat, and somehow get that heat away from you, i.e. away from the spacecraft you or your valuable payload are riding in. If you're traveling in orbit at 28,000 kph (17,000 mph), a typical pace in low Earth orbit, you have to dissipate 30 billion Joules for every ton you're bringing back. You could pop about a million bags of popcorn per ton with that.

Furthermore, the atmosphere is a windy, uncertain place. You have to get into more atmospheric science trying to make sense of it. After you make it through all that heat, the "winds aloft," as they're called, could send you all over the place. So if you're looking for a soft spot to land, and you aren't quite exactly sure where that's going to be, you choose the ocean. Florida being where it is makes it all the more reasonable as a place to land because the ocean is immediately adjacent to the launchpads. Cape Canaveral is right on the Atlantic Coast, after all. If engineers feel they need even more room, they choose the

Pacific Ocean. With a whole navy, you can recover spacecraft wherever they splash down. The Russian payloads land a tad harder on the steppes, the vast open land in the middle of Eurasia. The serious advantage of landing on land (*sic*) is that recovery crews can just drive there. But you need a big ole open space.

Whether they were launching and needed open space "downrange," or landing and needed open space to allow for targeting uncertainties, each superpower space program played the hand it was dealt. Each came up with its own best solutions to the rocket-energy challenges. The U.S. has Florida and an ocean next door along with a navy capable of operating almost anywhere. Russia and its contiguous states have an enormous tract of open land, without any buildings or even trees to bump against on the way down.

Without cold war constraints, engineers can be more methodical about optimizing their rocket strategy. There are benefits to launching the equator and landing in the ocean, so why not launch right on the equator from a place with no mountains or trees in the way? Well, a company called Sea Launch (how about that for marketing?) has hosted a great many launches from an enormous multi-hulled barge floating in the middle of the Pacific Ocean. The difficulty is that out at sea there are no rail lines or highways, so getting to the launch spot is a long haul. And weather rocks the boat, which affects timing, access, and safety. As a result, Sea Launch has proven to be a niche business providing launch service to organizations that want to send just the right payload, to just the right orbit, at just the right cost.

If you want to improve our access to space, and to all of the vast benefits that come with it, you need to think bigger. Floating launch platforms are cool and all, but the truly important new ideas in rocketry are those that deal directly with the energy involved in getting off the ground. Turn the page and I'll explain.

# 32

# BUILDING A BETTER ROCKET EQUATION

Whenever you fly a rocket, the stakes are high. One clogged fuel line, and the whole thing doesn't work; it explodes in midair or crashes. That is why rocket launches are so fantastically complicated and expensive. That is also why I am skeptical about high-frontier ideas like cooling off the planet with space-based sunshades, or surrounding Earth with a fleet of humongous solar-power-generating satellites. Even if those were good ideas (which I doubt), the cost and complexity of making them happen would be staggering. Rocketry is still just really hard. On the other hand, if we could make it better and more efficient there are a lot of other, more realistic things we could do to help humanity.

Okay, Bill, I hear you asking, you are the CEO of The Planetary Society: How do we build a better rocket? *Hmmm.* Not so easy. Basically you want to make it cheaper and more reliable. If you invest in enough rocket fuel to put your one-ton payload in orbit, you do not want anything to go wrong. Whether it's a telecommunications satellite, or you and your loved ones, you can't afford to blow things up. Just in the past few months, as I am writing this chapter, two supply ships to the International Space Station were lost; one blew up, the other failed to make orbit. A typical resupply mission to the space station has a payload of about two tons, and it costs around $125,000,000 per launch. There is a lot of room for improvement here. And since most of the rocket is fuel, that's also where most of the improvement will come: By making better use of energy.

It's worth noting that most rockets still use rocket fuel. I'm not kidding—it's called RP-1, Rocket Propellant #1. It's kerosene with chains of twelve to fifteen carbon atoms. It's very much like heating oil, only very well filtered and refined. There are no particulates or bits of paraffin to clog up your pricey, you-only-fly-once rocket fuel pumps. Another subtle but significant feature is that it's slippery, so it flows well. Since RP-1 is petroleum, it leaves a trail of carbon dioxide and water vapor. You have to think we can do better—and we can. The problem is not so much the pollution (although if we start launching a lot more rockets, it would be); no, the problem is that petroleum is just not perhaps the best rocket fuel. There are a few better ways to move a spacecraft.

A straightforward answer is to use liquid hydrogen. It contains more energy per kilogram than RP-1, and it produces little more than water vapor exhaust when it burns. It's also a proven technology. It powered the second and third stages of the Saturn V rocket that took the Apollo astronauts to the moon. The Space Shuttle also had a huge liquid hydrogen plus liquid oxygen–burning engine. In some not-hard-to-imagine future time when we fly a great many more rockets than we fly today, we might use electricity from wind and solar sources to split water, producing hydrogen and oxygen. We'd liquefy them with renewably powered refrigeration systems, and recombine them in high-powered rocket launches. This is the one place where a hydrogen-energy economy might actually make some sense. But in many cases, there are better solutions than hydrogen rockets, too.

Quite often, enthusiasts complain to me that there are no new ideas in space because "we still use chemical rocket fuels." Well, we actually have some technologies that were the stuff of science fiction when I was a kid that are flying on a great many satellites right now. The first is ion propulsion. An ion is an atom or molecule that carries an electric charge. That means you can move it with an electric field, which opens up a really interesting possibility. What spacecraft engineers do is rig up an electrical grid—a hi-tech metal mesh—at the back of a space probe and give it a negative charge using electricity provided by a set of solar panels. Inside the probe is a small tank of liquefied xenon, an inert gas that is especially easy to work with. The engineers use a valve to release a gentle stream of xenon gas, and then use electricity to strip virtually all of the electrons off of the atoms so they carry a strong positive charge. Then those charged ions (xenon atoms) go zipping rapidly toward the negatively charged grid.

Electrical attraction is powerful, so the xenon atoms shoot out of the probe moving fast, I mean really fast: about 50 kilometers per second, or 100,000 miles per hour. They fly right past the grid and shoot out into space. For every action there is an equal and op-posite reaction, so as they fly backward they impart a forward push to our spacecraft. Sounds wild, but it is real, and it works. The Dawn spacecraft, currently exploring the dwarf planet Ceres, uses this tech-nology. Little puffs of xenon ions accelerated Dawn to 10 kilometers per second, about 10 times the speed of a rifle bullet. Engineers man-aged to do this because the exhaust velocity is so high. Each ion con-tributes a tiny, almost unmeasurable push, but there are a lot of them and they are all moving at extraordinary speed. Better yet, an ion engine can run continuously for days or months. The spacecraft keeps building up speed, and it can steer as it goes.

I should reveal here that there is one big catch. An ion engine pro-duces a little thrust over a long time. That means it cannot get you off the ground; it works only once you are in space. You still need RP-1 or hydrogen or something for the initial launch. Once you are up there, though, an ion engine is far more efficient and persistent than any chemical rocket.

Another rocket-propulsion scheme of the future might involve making combustible materials that burn hotter than they would on their own by blasting the fuel with a superhot laser beam or maser (microwave amplification by stimulated emission of radiation) beam. The beam would come from equipment on the ground. We could pump extra energy into the engine from the outside. It's an exotic idea that seems to hold great promise. It could make launches cheaper and easier.

At The Planetary Society's Humans Orbiting Mars Workshop in the spring of 2015, someone remarked, "Getting to Mars would be so much easier if we had nuclear propulsion." Back in the 1960s, some visionaries really did imagine that we would explode "small" nuclear warheads behind a spring-loaded metal plate attached to a spaceship, and the pressure of the explosions would push the spacecraft along at higher and higher speeds. Someone would have to build the bombs and fly them up into space and explode one every few seconds behind a ship full of something valuable, like people. This almost certainly will never happen, at least not in my lifetime, but a nuclear rocket might make sense in other ways. There's an idea to use a superhot nuclear reactor to superheat liquid hydrogen. We wouldn't burn the hydrogen. Instead, it would expand as a gas and go shooting out the exhaust nozzle at crazy high speed, much faster than we can do with hydrogen and liquid oxygen combustion. A decade ago, NASA started work on a robotic (no humans aboard) mission to Jupiter using this technology. It may make a comeback.

My favorite exotic propulsion scheme involves a different kind of nuclear reaction, specifically fusion reactions, in which hydrogen nuclei become helium nuclei. I'm talking about the continuous, white hot, explosion that is our star, the Sun. It is surprising at first, but light pushes on things. Although light is pure energy and has no mass, its photons carry momentum. So if you (we) have a spacecraft of low enough mass and a big enough area for the Sun's photons to bounce off of, the spacecraft gets a push. This technology is called "solar sailing" because these spacecraft can be flown in a fashion quite similar to the way boats are sailed. You can run down-light (like downwind). You can tack. You can run across the light (across the wind, akin to a sailboat's broad reach). And you don't need to carry any fuel with you because the sun does all the work. It's a whole other way of tapping into solar energy.

I first heard about the idea of solar sailing when Carl Sagan was on *The Tonight Show* with Johnny Carson in 1976. Professor Sagan gave us an earful later that school year when I was in his astronomy class. The idea was to unfurl or deploy a super-shiny, very low mass sail in space. Then sunlight gives it a push. The 1970s concept was to build an enormous sail, a great big square, about a kilometer on a side. It would have been launched from Cape Canaveral. Its orbit would have precessed over the North and South poles. Then the spacecraft would harness the pressure of sunlight to sail way, way out and catch up with Comet Halley (aka Halley's Comet).

The Comet Halley mission was scrapped to free up funding for the Space Shuttle program. Nevertheless, the dream of flying a solar sail stuck with the JPL engineers. My predecessor Lou Friedman was one of the mission planners. He went on to write a textbook on solar sailing in the 1980s. I bought it and read it. I became the CEO of The Planetary Society, which was founded by Carl, Lou, and Bruce Murray (the head of JPL in the 1970s). And thirty-nine years after Professor Sagan showed the idea to the world on *The Tonight Show*, The Planetary Society pulled it off. We flew our LightSail in the spring of

2015. The sails deployed, and we got some remarkable images down. Our plan for 2016 is to get a ride on a NASA rocket to a higher altitude where we can really prove out the concept. We'll steer and tack and increase our orbital energy. It's a realization of a decades-old dream.

If you have enough time and patience, a solar sail can take your spacecraft to all kinds of significant destinations in the solar system. You could catch up with a comet, for example. As you may know, when an object is orbiting close to a planet or the Sun, it has to have a faster orbital speed than a similar object at a further distance. Venus orbits faster than Earth. If an object orbiting Venus were to try to keep pace with the more distant Earth, it would spiral into the Sun and disappear in superhot, catastrophic fashion. However, if we fitted a spacecraft with a large enough solar sail, it could orbit much closer to the Sun than Earth does, but nevertheless keep pace with Earth.

Ships at sea often practice what they call "station keeping," holding their positions relative to one another. This is the same idea in space. A spacecraft keeping pace with Earth could be fitted with instruments that would give us a few hours warning of a coronal mass ejection (CME) on the Sun, a huge explosion that produces high-energy particles that can damage power lines and communication satellites. It would be a remarkable mission that solar sails are uniquely qualified to carry out. In similar fashion, a sail spacecraft could be stationed to monitor Earth-orbit-crossing asteroids. Such objects need our watchful eyes. As we like to say, there is no evidence that the ancient dinosaurs had a space program. If they did have some good rockets and deflection technology, perhaps they wouldn't have been wiped out by an impactor from deep space.

Let's say we get a solar sail deployed in Earth orbit. Using gyroscope-style spinning momentum wheels, we can twist the spacecraft in inertial space. We have sunlight push the sail as it moves away from the Sun. Then we have the spacecraft twist so that it flies edge-on, like a karate chop as it flies toward the Sun. This is just what The Planetary Society will do with its LightSail spacecraft in the fall of 2016. It's a scheme that could greatly reduce the cost of exploring farther and deeper into space. As well as they might be suited to deep space missions, solar sails are not well suited to carrying people. Sails are for robotic missions. But a solar sail is clearly part of our future. Sails are like arrows in our exploration quiver.

To lower the cost of space exploration, there are still a great many problems to solve and technologies to pursue. Among these is the "beer can problem," another big opportunity for rocketry. The mass of beer compared with the mass of the metal in an aluminum beer can is about the same as the mass of the fuel in a rocket compared with the mass of the rocket itself. A rocket has to be almost all fuel. The way I see it, there's a lot yet to be done in materials science. The less the rocket weighs, the more mass is available for the payload. Comparing rockets to airplanes, about 10 percent of a plane's takeoff weight is fuel; a rocket's fuel comprises over 90 percent of its launch weight. Seems like a solvable problem. Rockets are made largely of metal—aluminum, titanium, and some stainless steel. If these materials could be replaced with carbon fiber, reinforced plastic, carbon nanotube fibers, or something we haven't thought of yet, we would greatly reduce the cost of doing things in space.

You might think, "Well, why don't we put wings on rockets and fly them up to the top of the atmosphere, where we could fire a rocket engine and really get going?" It's a rocketeer's dream. The problem still is that wings are heavy and it's a lot to carry up and down. Take a look at the old Space Shuttle. It landed with its wings all right, but it got into space strapped to the side of a rocket. The venerable X-15 rocket plane was carried aloft by a B-52 bomber. The Virgin Galactic Space Ship 2, designed to carry space tourists, is largely carbon-fiber-reinforced plastic. It's very lightweight compared with traditional metal rockets. But it is nevertheless carried to a high altitude by an airplane, also made of lightweight materials. In the Virgin Galactic scheme, there is an airplane whose energy comes from aviation gas, and there is a rocket engine that burns rocket fuel. It's a good system, but still it has two vehicles, an airplane plus an aircraft-spacecraft that becomes an aircraft—two systems to get into space. Same with Microsoft billionaire Paul Allen's Stratolaunch Systems. The plan is to have a plane carry a rocket to a suitable altitude, where it lets it go off into space. Meanwhile, for over fifteen years, Orbital Sciences Corporation has done the same sort of thing with the Pegasus rocket. It uses winged lower stages to get a rocket stage high enough to scoot into space.

Competing with Virgin Galactic et al is X-Cor and its long-in-development Lynx space/aircraft. It looks like X-Cor's Lynx spacecraft may indeed live the runway-to-space dream. It's made of plastic and carbon-fiber composites, and it's powered by rocket fuel from the runway, up into space, and back. We'll soon see if this idea is viable. Keep in mind that both of these carbon-fiber air/spacecraft take payloads (humans and their cameras) only to the edge of space. They do not have nearly enough energy to get in or on orbit. But I feel they're taking us closer than ever. It sure looks possible to one day leave a runway, fly into orbit, reenter, and fly back to Earth landing on a conventional runway. If we simply had the materials that were just a little stronger and a little lighter weight, it looks like we'd pull it off.

As I mentioned earlier, space brings out the best in us. That's why the Indian Space Research Organisation (ISRO) flew and successfully put a spacecraft in orbit around Mars, an extraordinary mission. For a relatively small investment of about $70 million, the Indian government advanced its aerospace technology and inspired a whole generation of budding scientists and engineers. That's why South Africa has a space program to put its citizens in touch with orbiting spacecraft and build a countrywide array of radio telescope dishes. Mexico and

Britain each just rekindled their space programs. Canada is so proud of its plucky Space Agency that their Canadarm (a robotic attachment to the International Space Station) is featured on the back of Canadian five-dollar bills. The relatively small country of Vietnam has a space program. Citizens around the world want to bring out the best in themselves, and space is the place for that.

So stay tuned, my fellow Terrans. The best is yet ahead for us in space and back down here on Earth. We will continue to make extraordinary discoveries out there in deep space. And these new discoveries will be made not by an individual like Copernicus, Galileo, Kepler, or Goddard, but by our entire society. By understanding the rocket science involved, by mastering another form of energy, we will explore farther and deeper into space. We will search for signs of water and life on other bodies in our solar system. We will explore other solar systems. We will learn more about Earth and about ourselves in the process, and we will nurture the future visionaries who will keep making our society better. Rocket science has changed the world before. It can do so again.

# 33

# DO HUMANS HAVE A DESTINY IN SPACE?

While we are working hard here on Earth to address climate change, I feel strongly that we must continue to advance space science and space exploration. I can tick off a bunch of practical reasons why the two go hand in hand; I described some of them in the previous couple of chapters. But to me there is also a bigger idea at work. It comes back to the whole notion of what I think it means to be a Great Generation. Greatness is about more than facing up to a huge challenge—climate change, right now—and finding ways to overcome it. Greatness has to include exploration and expanding the most noble part of our human spirit, our ability to know the cosmos and our place within it, to experience the beauty and majesty of the universe, our ability to achieve a peaceful future by working together. It's not that we won't compete; it's that we'll compete toward this common goal—to leave the world better than we found it.

There are two questions that get to us all. If you meet someone who claims that she or he has never asked these questions, she or he is almost surely lying, to you or himself or herself. Or maybe you're dealing with an unsophisticated robot? Here are the big questions: Where did we come from? And are we alone in the universe? Putting these another way: How did life begin? And is there another intelligent life-form out there? Can we be in touch with it, or them? If you want to answer those questions—and I claim that we all do—you have to explore outer space.

Every civilization and tribe on Earth has a creation story, a myth that explains how we all came to be here. In the world of modern science, we have one, too. We've studied the cosmos carefully and come to realize that there must have been a primordial moment, an instant of time $10^{-42}$ times briefer than the blink of an eye, when the universe was concentrated into a volume far smaller than an atom. It exploded, for lack of a better word, and all that we can observe in nature, including ourselves, emerged from that Big Bang. Furthermore, it's clear that we are made of the very same material as the distant stars. We are stardust, built from elements that were forged in the nuclear furnaces of ancient supernovae.

For me the most astonishing part of all is our consciousness. We can know that we are stardust; we can know the universe and our place within it. That insight would not be possible without space exploration. At the heart of the challenge of climate change is being a good caretaker of the planet—of viewing Earth as both house and home. Space exploration is where that spirit comes from. It is the greatest manifestation of perspective, and it is the kind of stretching that gives higher meaning to all of those other daunting everyday challenges.

When I was in elementary school, no one, my second-grade teacher Mrs. McGonagle included, knew what became of the ancient dinosaurs. In my lifetime, the worldwide layer of iridium metal and a remnant impact crater off of what is now Chixalub, Mexico, were discovered. So far, these are the best evidence we have for an impact so energetic that it produced heat sufficient to set the world on fire, while at the same time creating a global dust cloud that blocked the Sun and produced catastrophic climate change. The more we study the solar system, the more reasonable this scenario becomes and the more aware the scientific community especially becomes of the potential for unimaginable danger and hardship.

With this relatively recent realization, we also become aware that we are the first generation of humans that could do something about an incoming asteroid or comet. Knowledge gives us power over our destiny, just as is true with climate change. Based on telescope observations of the solar system, astronomers now estimate that there are at least one hundred thousand Earth orbit–crossing asteroids comparable in size to the one that finished off the ancient dinosaurs. If any one of these hit our planet this afternoon, that would be the end of everything we know. We as a society have an opportunity. We could direct a small fraction of our intellect and treasure to identify the dangerous objects and then build a spacecraft capable of nudging one of these things safely off of a collision course. I'm talking about giving certain line items priority in, say, the NASA, European Space Agency (ESA), Roscosmos, China National Space Agency, and JAXA (Japanese Exploration Agency) budgets. Detecting every single seriously dangerous object out there is perhaps a billion-dollar project. Put another way, it would cost the amount of money that the United States government spends every two hours. A two-hour investment could save all of humankind from the most unpleasant form of global change.

Other objects in the solar system, including the Moon and Mars, get hit by asteroids and comets all the time, too. In learning about the dangers of impactors, we are also learning amazing things about the history and origin of the solar system. The Moon as a whole is a product of the infant Earth getting hit with a smaller, but still planet-size, rock going very fast in ancient outer space. Mars has been blasted by impacts so powerful that they threw bits of the planet's surface up into space, where some of them landed here on Earth as meteorites. The ice sheets of Antarctica act as huge meteorite collectors, scooping up fallen rocks and depositing them in convenient groupings. If you know where to look on the snowy ice near the Allan Hills in Antarctica, you can find rocks that were knocked loose from Mars by an asteroid that struck about 3 billion years ago.

At the time when Mars got smacked, its surface was quite wet. We know this in part because we can still see traces of the ancient water in those Mars meteorites. Our space probes also observe traces of that long-lost warm climate. Back then, the Red Planet had rivers, and lakes, and bodies of water everywhere. Since then, Mars's smaller size meant that it did not develop and maintain a spinning iron core that would produce a magnetic field similar to the one we have here on our Earth. So the solar wind, the stream of particles emanating from the Sun, slowly stripped a great fraction of the Martian atmosphere away. But what if life started on Mars during that balmy age? Even more startling, what if Earth life actually started out on Mars? Microbes might have stowed away in the pit pocket of a Mars rock that got blasted to space, just like the rocks we find in Antarctica. Perhaps it landed on Earth and found a friendly new home. Perhaps it settled in and reproduced. Perhaps you and I, and every creature we know, are actually descendants of Martian life. Whoa . . .

If we found microbes on Mars that are clearly related to those on Earth, such a discovery would change the course of human history. It would be as stunning as Copernicus's realization that Earth orbits the Sun, rather than the other way round. It would be like Galileo discovering that the Moon is imperfect, that Jupiter has moons of its own, that the Milky Way is made of innumerable stars. In the same way those earlier discoveries did not change everything overnight, neither would or will the discovery of evidence for life on Mars. People would

still drive on the right in the U.S. and on the left in Japan. But everyone everywhere would soon come to feel differently about what it
means to be a living thing in the cosmos. It would, I expect, give us
profound reverence for our unique planet and our amazingly comfortable place in space.

Mars today is not a hospitable planet. It is amazingly cold all the
time. Forty below is a typical noontime high (−40 is the same in Fahrenheit or Celsius). There is hardly any water anywhere, and there is
definitely no liquid water on the surface; the air pressure is so low that
it would immediately boil away. What atmosphere there is consists
mostly of carbon dioxide, and the surface pressure is just 0.7% what
it is on Earth. At the same time, much of the landscape on Mars looks
eerily familiar. Mars shows what happens to a nominally similar planet
that does not have Earth's special, life-preserving qualities. Mars is too
small to preserve a protective magnetic field and to hold on to a thick,
warming atmosphere. As a result, it lost most of its greenhouse effect
and lost most of its water. When we look at other worlds, we gain perspective on our own.

So despite the harsh environment, we must explore Mars. The
potential for making an extraordinary discovery is just too high to
ignore. A hundred years from now, people will probably know whether
or not there was or even is life on Mars. A hundred years from now
people will probably know, too, whether there is life on Europa, the
moon of Jupiter, which has twice as much seawater as we do on Earth.
The search will continue at Enceladus, a moon of Saturn that has a
warm, salty ocean the size of Lake Superior under its icy south pole.
Someday we may send a submarine to Titan, another Saturnian moon,
where the surface is dotted with lakes made of liquid natural gas
and where the chemistry may resemble a deep-freeze version of early
Earth's. For my part, I would much prefer that we set about exploring

and knowing these other worlds today, in my lifetime. We know how to do it. We even have the plans ready to go. It just takes the commitment and not even an especially big one.

Commitment is a rare thing, however. We are probably not going to have another "Kennedy moment," the speech given by President John F. Kennedy at Rice University in Texas, in the U.S. on 12 September 1962. Kennedy said that we would go to the Moon and accomplish the necessary technical feats along the way, ". . . not because they are easy, but because they are hard." The actual reason is found a few moments later in the same speech, when the president remarked, ". . . that challenge is one that we will accept . . . and one that we intend to win." He was talking about beating the Soviet Union to the Moon, because it would demonstrate the U.S.'s technical capability. A year after Kennedy was assassinated, the U.S. Congress authorized a growing NASA budget that eventually reached over 4% of the entire U.S. federal budget. That money allowed NASA's engineers and technicians to create a human space program and train pilots to make the trip. That will almost certainly never happen again.

Today's NASA, like all the other space agencies around the world, has to compete quietly with the hundreds of other line items in the federal budget. Any exploring we do will have to occur within tight financial constraints, without any of the old special priority. As I write, NASA's budget for planetary science and exploration is a little less than $1.5 billion a year. That's nothing like 4% of the federal budget. In fact, it is just about 0.4% of the budget. Even with that, we can do astonishing things. It is enough to send rovers rolling on Mars, and to get pictures coming back from Mercury, Saturn, and perhaps the most extraordinary place of all, Pluto.

No humans go along on any of these missions. Right now, no humans are going anywhere more than 400 kilometers above Earth, where the International Space Station orbits. Having humans out there exploring will have great value. The scientists and engineers I work with have shown that what our very best robots—designed by our best robot builders, directed by our best scientists—can do in a week on Mars, a properly equipped human explorer could do in about fifteen minutes. But getting a human out to Mars and back is and will be extraordinarily difficult, much, much harder than a round-trip to the Moon.

After careful analysis, we at The Planetary Society are advocating strongly for a journey on which humans would orbit Mars in 2033. You can go to Mars only every twenty-six months or so. At other times, the planets are just too far apart, on opposite sides of the Sun. We have shown that the NASA budget would not have to be expanded to support such a mission; no second Kennedy moment would be required. With routine budget adjustments for inflation, NASA could have people out there in orbit around Mars in less than twenty years. Subsequent missions would take astronauts down to the Martian surface. Such missions would be extraordinary. They would demand the best from us. People all over this world would follow very closely what was happening out there on that other world. When we go exploring like this, two things happen. We make discoveries, to be sure. But we also have a grand adventure. Humankind would be fully engaged.

Elon Musk, the man who got rich by inventing and investing in companies that became the original practical Internet purchasing service, PayPal, dreams of going to Mars much sooner than 2033. If NASA is spending just $1.5 billion on the planets, and with ESA, JAXA, and Roscosmos spending a fraction of that, it's easy to imagine what a driven multibillionaire might be able to accomplish. But Mars has proven to be especially hard, and I think governments working together have by far the best shot at getting humans there. Just this morning, as I write, a SpaceX Falcon-9 rocket exploded a few minutes after leaving the launchpad. Fortunately it was carrying only cargo for the International Space Station, but still . . . In the space community, we all wish SpaceX the best as they work their way back from this expensive failure. Meanwhile, there are many other companies pushing forward with both crewed and un-crewed rockets. We'll see what becomes of SpaceX, the United Launch Alliance, Sierra Nevada, Blue Origin, Orbital Sciences, and their competitors in the coming years. I hope it is very, very exciting.

It's easy for me to imagine how a Mars mission might go. We find a super-salty layer of ice that is exposed on one edge in a steep gully; we've already seen hints of them from our robotic probes. Every Martian year that ice gets warmed to slush. And when we dig into that slush, we find that there is something still alive—a microorganism, a "marscrobe" of some sort. I hope we find evidence of life on Mars in my lifetime. I will want to know whether or not it is like us. Do marscrobes have DNA similar to our own? Or are they something completely different? Either way the discovery would be astonishing and profound. If they are completely different, it would suggest that life is pretty common in the universe. If they have DNA and similar life processes, it would suggest that we, all of us, oak trees, giant squid, poison ivy, phytoplankton, and my old boss are descendants of the ancestors of those marscrobes.

Imagine the impact of such a discovery on medical science, let alone on biology, sociology, chemistry, and physics. Talk about a world-changing discovery. Cue the spooky music, and let's explore.

The excitement and profound nature of space exploration is why I accepted the job as the Chief Executive Officer of The Planetary Society in September of 2005. I am a charter member. Space exploration brings out the best in us, and from a strictly practical standpoint, our ability to navigate spacecraft may literally save our world in the coming years. Earlier I mentioned that the Society was founded in

1980 by Carl Sagan, Bruce Murray (the head of JPL, the Jet Propulsion Lab), and Lou Friedman (a JPL engineer). They felt that public interest in space was very high, while government support was waning.

In an important sense, the same thing is happening now. The budget numbers are not encouraging. And yet I have also seen the outpouring of joy and enthusiasm when the New Horizons mission flew by Pluto and returned the first images of this distant world, and when The Planetary Society announced its LightSail mission. For me, the response is incontrovertible evidence that people everywhere want to support spaceflight. The Planetary Society is by no means the only private organization helping to revive our species' explorations of space. There are other companies doing it, too. B612 is working on a private mission to find dangerous asteroids. Deep Space Industries and other organizations want to mine asteroids. Several companies are gearing up to take tourists on brief journeys into the blackness of space. People just want to get up and out to see what it's like to be an astronaut and peer down at our world from far above. We are living at a turning point, an extraordinary time for humankind in space.

Before we conclude here, just consider how much exploration means to our species. We naturally go over the next hill to see what's beyond the horizon. Our ancestors who did not feel that compunction did not actually become our ancestors. They got outcompeted by individuals and tribes that did explore. Look at it this way: What if we stop exploring space? What if we stop looking up and out? What would that say about us? Whatever it is, it would not be good. So as we address climate change, let's continue to explore space. Let's draw inspiration from all its great challenges. And let's use those challenges to foster international collaboration, so we bring together the greatest minds from around the world.

Although The Planetary Society is an international organization, it is based in the U.S. We keep an eye on NASA and we routinely lobby representatives in the U.S. Congress, because NASA is still the world's premier space agency. It is currently constrained by political forces to build hardware and support infrastructure and systems for murky missions to destinations in the solar system including, allegedly, Mars. The rockets and deep-space habitats that are being built and planned are what they call "mission agnostic." In other words, they're building astonishingly expensive stuff without a clear idea of where they're going to send it. This strategy, or lack of one, has turned some parts of the world's most inspirational organization—the best brand the United States has—into a make-work program.

Here's hoping in the coming years we can turn this around and have NASA lead an international effort to send humans to Mars, develop the new technologies required, make amazing discoveries, and make everybody's lives better in the process.

# 34

# SETTING A FAIR PRICE FOR A BETTER PLANET

As an engineer, I pretty much can't help it: I imagine a technical solution to pretty much any problem you can think of. Can't find your luggage? Put a bright orange tag on the handle. Cars are going too fast? Reshape the street with speed bumps. Hungry? Design a small raised-bed vegetable garden out back. Hot water takes too long to get here? Let's plumb a pump under the sink. It's not until the next step, when a technical solution doesn't work, that I even consider changing my behavior. Cars are going too fast? Okay, maybe, just maybe, if the speed bumps don't work, and if there is no way to install a variable speed limiter on the car or replace it with a computer-controlled car or to reprogram the navigation system to reroute cars away from the neighborhood where they are going to fast . . . maybe then, I'll look at behavior. We could pay sufficient taxes to put a police officer on that route and enforce the speed limit . . . I suppose.

But when it comes to climate change, along with what I believe are going to be the huge energy production, transmission, and storage projects that we will need, we are going to need effective ways to discourage people from pumping out greenhouse gases, especially carbon dioxide. I'm talking about regulation. I'm talking about harnessing the kind of power that tinkerers might not talk about: the power of economics.

I know, I know, the conservatives and the libertarians among us just do not cotton to taxes of any sort. My impression is that conservatives right now equate taxes with evil. They may be right; taxes may be evil. If that is true, I cannot help but observe that every government in the world makes extensive use of taxes, so I guess it's further evidence that Satan is real and he has taken over. Seriously, though, without taxes you have no collective order, and without order it's virtually impossible to achieve the time-honored "rule of law." It's hard to believe that anybody genuinely thinks that a tax-less society would have much of a future. So how about we consider a carbon "fee," instead of tax? A fee doesn't apply to you unless you choose to partake of a commodity or make use of a service. No taxes; just fees.

Keep in mind that Engineer Bill hopes that technical solutions will emerge on their own. That's why I wrote most of this book mapping out ways it could happen. As entrepreneurs see the market for a home battery system that can power a house overnight, those battery systems will come on the market. This should happen even if it means nothing especially new in the chemistry or function of the batteries that become the new electrical-storage product. They'll just be existing batteries incorporated in new good-looking packages at first. This market will encourage other people to innovate, and especially, to invest. I like to think of this as organic, natural growth of a new market. New things will just happen because of what consumers, industry, and power companies want.

But when it comes to greenhouse emissions, the market forces aren't being allowed to work as they could or should. Right now, no one pays for the release of carbon dioxide, methane, and other greenhouse gases. Rather than saying "carbon dioxide," it's popular now to just say "carbon." No one pays for carbon. Everyone who produces it just goes on her and his merry way. Every time we drive anywhere. Every time we use plastic bags for groceries or dry cleaning. Every time we fly in a plane for vacation or business, we are putting the waste gases, the waste carbon into the atmosphere and not paying for it. This has been true since Og and Ogette were sprinting, ambling about, hunting, gathering, and scavenging on the ancient savannah. We've built fires using whatever fuel we could find—to keep warm, to see in the dark, to cook, and to discourage predators since the dawn of humankind. Our effect or impact on the environment was largely minimal, until we really revved up burning things around 1750 with the invention of an excellent steam engine.

Even as carbon emissions kept increasing, nobody owned up to their cost. That's been true for companies as well as for individuals. Along with contributing to our economic well-being, industrial activity contributes to our troubles, producing every bit as much greenhouse gas emission as all of our transportation or all of our agriculture. Business managers can figure out how to be more efficient. They understand and assess the economics of everything they do. When oil prices shot up in the 1980s, for instance, the entire U.S. economy became radically more efficient (during a period of strong economic growth, no less). Go into any hotel today, and your way is lit with energy-efficient lightbulbs. Hotel managers changed the bulbs because it saves the business money. If we continue with the practice of free carbon dumping, we're headed for trouble. If everyone has to collectively (and invisibly) pay for damage to our environment, companies and industries are going to keep on going as they have the last couple of hundred years. It's time to make the big change. Companies are not generally malicious any more than you or I. If I offer you something for free, is it your fault for taking it?

I am pretty sure that situation is going to change, not only because it will have to, but because that is the fair way to do business.
Economists are fond of coming up with mathematical models to describe human behavior in trade and commerce. In economic terms,
carbon dioxide exhausted into the atmosphere is an example of an
"externality," or "externalized cost." What they mean is, the people
who make the carbon do not pay for it. Instead, we all do. Everyone in
the world pays for everyone's carbon production, because we have
one atmosphere. We all share the air.

Two classic, oft-cited examples of externalized costs and consumers are vegetables in Pennsylvania and T-shirts and toys from Asia. On
the east coast of North America, especially off-season, in the wintertime, you can buy vegetables that are very fresh, because they've been
shipped from the fertile valleys of California on the west coast of
North America. They arrive in refrigerated trucks and trains. Those
vehicles and railroads burn fossil fuels to move all this food. Carbon
is put into the atmosphere by the kiloton on each trip, and no one at
the food corporation or the retail grocery store or market pays for that,
at least not directly. When goods are manufactured by low-wage earn-
ers in the developing world, they are put "on the water," as the saying
goes, and shipped to North America aboard vessels that spew megatons of carbon into the air, externalizing the shipping cost. But if you
think about it for just a moment, what is really happening is that we
all end up paying for the carbon dumped into the air.

If one is looking for a scapegoat, or perhaps a trip (a herd) of
scapegoats, to blame for this externalizing cost feature of our economy, the most popular choice is "corporations." It's easy enough to as-
sert that it is The Man who is causing climate change. It is a shortcut
to believe that corporate pigs (or goats, the scape kind) worldwide are
out to ruin your life by their blatant disregard for all living things in
general and all that is sacred. But really, this is not much more sensible

than declaring that taxes are evil. It seems to me that corporations are just behaving rationally. Their managers feel they generally are doing the right thing and being perfectly moral. Nobody thought much about the impact of carbon emissions until pretty recent times in human history.

Having explained it this way, I hope we can agree that what we need to do is internalize the cost of $CO_2$. We have to find a way to induce producers of greenhouse gases to pay for their share. People around the world have readily agreed on this idea in principle—and then fought about it like rabid, angry, disenchanted cats and dogs!

You may have heard of the Kyoto Protocol from 1992, which was intended to extend the UNFCCC, the United Nations Framework Convention on Climate Change. That was just the beginning of the process. Then there was the Hague Climate Conference in 2000, often called "COP6" (the sixth session). The Copenhagen CNFCCC in 2009. Doha in 2012. Did you hear about them? Perhaps you thought hard about heading to Bonn and Paris conferences in 2015. At these conferences technical people and politicians just agonize over who ought to do what about climate change. In general, no one has the guts or the political influence sufficient to establish a carbon fee or tax or shared financial burden regarding climate change. Oh yes, "nonbinding" agreements are reached.

Government representatives all very much want to do the right thing. And they parse every detail, from the number of trees grown here or there to the number of grams of carbon spewed there or here. But in almost thirty years, not much has changed. By way of example, on the UNFCCC list of accomplishments are diplomatically agreed upon bullet points that start with these phrases: "Strengthened their resolve"; "Streamlined the negotiations"; "Emphasized the need to increase their ambition"; "Launched a new commitment"; "Made further progress toward establishing the financial and technology support."

Speaking plainly, as a professional engineer and Emmy Award–winning writer whose favorite book of all time is still *The Elements of Style*, by Strunk and White, and also as a regular person very familiar with English, I must point out that these words mean almost nothing. I'll acknowledge, however, that these phrases mean a little bit. They indicate everyone's desire to do the right thing. I've been to one of these conferences. People are working hard, trying to get along, but hardly anything has gotten done on a global scale to reduce carbon emissions on a global scale. And here's why: My country, the United States. We use 18 percent of the world's energy, pump out 19 percent of the world's carbon dioxide, and we have only 4 percent of the world's population. That's it. Without the U.S. in the lead, ain't no nothing going to get done on climate change (employing the triple negative for emphasis).

It is no wonder representatives, negotiators, and diplomats from around the world are often completely frustrated with my beloved home country. My claim is simply that if the United States were doing more than nibbling around the edge of the problem, if the U.S. were leading in every technical and policy-related means to address climate change, the world would follow suit. We would be getting it done. In 2015 a nominal agreement was reached between the two biggest producers, the U.S. and China. Both countries promised each other to reduce greenhouse gas emissions quite a bit. President Obama issued a strong initiative regarding the reduction of coal burning. The U.S. agreed to cut 26 percent by 2025. China agreed to not increase its emissions after 2030, while also changing its infrastructure so that the country produces 20 percent of its energy from renewable sources. That's a start, but it's not enough, not by a long shot.

You've probably heard the term "cap and trade." The idea seems worthy. Every country would have a cap or ceiling or maximum amount of carbon and other greenhouse gases that each was permitted to pump into the atmosphere. Those countries that just can't get to the agreed upon cap would trade with other countries that might have a little credit below their cap to trade. It would be fine, if the situation weren't inherently complex and without a leader. In my opinion, each country's diplomats, generally at the behest of their bosses, have managed to get concessions from the United States, Russia, et al. But those cap-and-trade concessions are not getting the dumping of carbon into the atmosphere reduced fast enough. Cap and trade has not worked well enough, at least so far, because there are enough countries with carbon credit to spare that the whole scheme has been rendered ineffective. We need a different approach: something fair, politically feasible, and powerful enough to put a big dent in U.S. emissions. If we can do that, we can move the whole world with us.

Recently, I recorded scenes for a television special about climate change denial for National Geographic. As the host, I had the extreme pleasure of working with fellow actor Arnold Schwarzenegger. He is a consummate professional and a very thoughtful politician, especially with respect to climate change. In his office, he has busts of Presidents John Kennedy, Abraham Lincoln, and Ronald Reagan. It's a diverse bunch. (He also has a bust of himself.) He pointed out repeatedly that those guys were all true leaders. "Winners," he called them. We talked for quite a while about the success he had as governor in California establishing a workable cap-and-trade system to meet greenhouse gas and pollution emissions targets. The trades generally take place between states. California has an enormous economy compared with those of other states, so some trading is possible.

The former governor also pointed out that after progressive environmental laws are passed in California, succeeding governors do not undo what the previous administration accomplished. The regulations generally stay in place. In the U.S. at the federal level, this is often not the case. President Barack Obama has recently initiated some thoughtful, achievable environmental standards for reducing vehicle and industrial carbon emissions. We can't be sure that they'll stay in place after a new president enters the White House in 2017.

So here's a big idea, a huge and potentially world-changing idea, something that could be written into the law of this land that would change the world. The United States could establish a carbon fee and dividend system. (Not to worry: it's not a tax; that would be evil.) Whenever you or the corporation you work for or own creates carbon dioxide, you have to pay a fee. We would start out at, let's say, ten dollars per ton of carbon dioxide. A ton is a lot of gaseous carbon. And right now, ten dollars is about two fancy cups of coffee . . . per ton. That ain't much. That money would be directed into a fund that could become a trust fund akin to the Highway Trust Fund, established under President Eisenhower (a Republican, you might note) in 1956. Once the money is in that fund, it has to be used for its intended purpose—repairing roads, building mass transit, and cleaning up transportation-related storage tanks.

Congress has from time to time chosen to impoverish the high-way fund as a means to cut taxes (and ensure an increase in highway deaths, by effectively cutting maintenance). But so far, the fund still exists. It's tied to a tax on gasoline, which is apparently a popular thing to oppose. At any rate, once the money is in the fund, no official is allowed to do anything else with it. You can presume that there may be corruption, and so on. The concern seems to be that a small portion of the fund is often used for mass transit and transportation research and development, i.e. things that some people believe are not strictly roads, just infrastructure to support roads. But I don't see a strong ar-gument that there would be any more or less corruption than with any other U.S. activity.

We could use the fee money to build energy infrastructure, and the place to invest is in renewable-energy infrastructure. We could build a better transmission grid. We could invest in battery technology, concrete gravity pistons, solar photoelectric systems, solar hot water systems, better lightbulbs, and so on and on. Or it's been proposed that we just give that money back to every citizen. Everybody in the U.S. would get a dividend that could be used to offset our existing taxes.

This is a slightly counterintuitive idea (just slightly). The carbon fee would raise the cost of the things you buy (since right now there is some carbon emitted in the production and distribution of pretty much everything). That's a little less money in your pocket. But at the end of the year, the government would take all of the money collected by the carbon fee, divide it up, and give it back to you as a dividend check. By you, of course, I mean all of you. The government wouldn't keep any of the money. All the fee would do is put a realistic price on the carbon we dump into the environment. Every factory, every company would have an incentive to reduce emissions, because then they could sell things at a lower price. Consumers, given a choice between a low-carbon pair of jeans and a high-carbon pair of jeans, would see a cost advantage in choosing the former. If you live a low-carbon lifestyle all year, when your dividend check arrives you will find that you came out ahead.

For a small-scale example, I got into a good-natured competition with my neighbor Ed Begley Jr. over who could be more environmentally correct. Fee and dividend would encourage that kind of attitude across the country. As long as you use less carbon than the average American, you come out ahead. Everyone has a chance to compete.

Anyway I look at it, a fee and dividend system seems like a good idea, but it is absolutely a top-down arrangement. The government would be stepping in to collect and distribute this wealth. By the way, this infrastructure exists. Oh yes, The Man is already on your back.

We have the Internal Revenue Service, which provides the means to have the Antero Reservoir dam, air traffic control, safe food, clean water, and a military presence worldwide.

I imagine a great many of you reading along here think that such a policy is harebrained, unworkable, and perhaps even evil. Those of you who might think it's a charming idea also may see it as completely unworkable. Who would determine how much carbon each of us makes? Who would say how much carbon a ship produces compared with, say, a tire manufacturing plant? Well, gentle readers, we already do it all the time, day and night 24-7, 365¼. It's all science that we understand and know how to apply.

Whenever you or a corporation buys fuel, we know how much of that liquid fossil carbon will become a greenhouse gas. Whether it's your compact car on your short commute, or your fleet of immense container ships at sea, you're buying fuel. So the carbon fee is on the fuel. It reminds me of a utility sewer tax. In general, utility companies do not measure how much sewage you and your family produce. Instead, they measure how much potable or otherwise water you buy. Most of that, whether it's for taking showers, washing dishes, or making home-brewed beer, ends up going down the drain . . . into the sewer system. So that's what they tax.

Notice that with this scheme, if your car gets more miles or kilometers per gallon or liter, you pay fewer pennies per mile or kilometer. Your car's efficiency pays off. There are reasonable arguments that a fee on carbon would affect the poor more than the rich. That is almost certainly not true. Rich people use more gas, heat bigger homes, own multiple homes, and buy many more airplane tickets than middle-class and poor people. Rich people would almost certainly pay more in carbon fees than anyone else—and note well, they can afford it. The dividend part of the fee and dividend can be inherently fair.

If we had this carbon-fee system in the United States, it would affect everyone in the world. Goods that came from Asia to America would be effectively paying a fee, because the ship would have to pay a fee for its fuel. That cost would be passed on to everyone who buys the goods to be sure. But it would be passed on fairly. In this example, companies in the countries buying the theretofore-shipped goods might find that it's cheaper, all things accounted for, to manufacture some of those goods at home. That would stimulate the local economies and reduce greenhouse gas emissions. For you economics buffs, it would internalize costs.

People and corporations alike would be motivated to reduce carbon emissions. Energy prices would more accurately reflect the societal costs that come with them, which would encourage the move to things like wind and solar. Farmers would be able to see the precise upside of becoming part of the renewable-energy economy. Factories would see a clear economic benefit in improving efficiency or phasing out dirty production processes. The builders that manufacture huge container ships would be strongly motivated to make their ships more efficient. Come to think of it, the Navy would shop for ships and planes that use less fuel right along with every individual and corporation. Let me throw in one more idea so crazy it just might work: What if those ships also produced the remarkable microbubbles that would increase the Earth's albedo and reduce global warming? And those shipping companies would get credit for doing that at the same time they're lowering the carbon fee and the cost to their bosses, shareholders, and the consumers who buy their products. It could be the new Wild, Wild Yes (*sic*).

Right now in the United States, along with the climate-change deniers, we have a lot of conservative politicians and commentators who just don't see how something like this carbon fee could possibly work (even though similar systems have been endorsed by conservative politicians in the not-so-distant past; there are Web sites devoted to their quotes). It is my perception that they see climate change as a wholly intractable problem. They seem to perceive addressing climate change as being like a game of Whack-A-Mole. You smack one mole down, and another mole pops up somewhere else on the game board. To me, that's just giving up before you get started. I don't know how much time you may have put in playing Whack-A-Mole, but I can assure you that if you spend the coins and put in the hours, you can get better at it. To whack those moles, you have to clear your mind, in the same way a good athlete does (or so I've heard).

Here's a good example of what *not* to do with respect to climate change. Several major cities around the world have implemented restrictions on when you can drive in the city. The practice is generally called Road Space Rationing. The idea is that since roads are shared, we should all take turns using them. So in places such as London, Paris, Mexico City, Santiago, São Paulo, Bogotá, Beijing, Quito, La Paz, and Athens laws have been passed that permit only cars with certain license plate numbers to enter the city on certain days. Some cities have tried permitting only odd or even license plates on every other day, for example. In general, these schemes don't work. For instance, people will go to the expense of buying a second car to have a second plate number so they can drive more. And, as though I have to tell you, those second cars are generally cheaper and less efficient than the original or primary car. Any solution that encourages people to game the system is not much of a solution at all.

Road Space Rationing is an example of a top-down regulation that imposes a government restriction that just doesn't work. But it absolutely does not mean that no scheme could work, or that we should not restrict inefficient use of our right-of-ways. It just means we need something better. The carbon fee would be a start. If your car is inefficient, you would be paying a higher carbon fee. But how about this: We could make public transportation systems so good that people come to prefer them. People going to and from work during peak hours would end up using the subway, the bus on its own buses-only lane, the protected bicycles-only roadways, or the sky train monorail (or Hyperloop) of the future. If the United States were in the business of promoting the best subway train cars, rails, and control and coordination systems for subway trains in the world, other countries would embrace them and use them, which would reduce greenhouse gas emissions worldwide.

In addition to improving the physical commuting that workers do, a carbon fee would also do a lot to encourage the kind of telecommuting that futurists have been predicting for . . . what fifty years? Telecommuting is finally catching on, a little. If the costs of traveling into an office every day were spelled out more clearly in carbon-fee dollars and cents, I'm pretty sure a lot more people would opt to work electronically instead. And yes, those electrons could come from renewable energy. The Google computers that search the Internet could be more efficient, the homes where the telecommuters work could be more efficient, maybe with a solar hot water heater on the roof. It keeps going and going. Once you set this change in motion, it will build and build on itself.

Say what you will; complain all you want about how nothing is made in the U.S. anymore, etc. But the one thing that we do export very, very well is our culture. People around the world learn English by watching all of our crazy and not-so-crazy programming. They watch our movies. They embrace our technological innovations. If the U.S. were leading, the world would be following.

Two close friends of mine from high school moved all the way across the North American continent to Alaska. Barclay met a guy; Ken met a woman. Each has lived there for the last thirty-five years. In Alaska, the government shares the wealth from the North Slope oil and gas. Everyone in the state gets a dividend every year from the

Alaska Permanent Fund. Right now, it's about two thousand dollars a year. You just have to live there; you can't be a registered voter living somewhere else. In 1976, people agreed that if oil companies stand to benefit in big ways from the fossil fuel beneath their land and their nearby sea, everyone should benefit. Alaska is not exactly a stronghold of the liberal left. Very much the opposite, by any measure. Nevertheless, people there saw the benefit and the inherent fairness of sharing the wealth. It just seems to me that if Alaskans—as hard-core libertarian as they may be—can see sharing wealth as being for the common good, perhaps we all can.

I cannot help but remind us all that people are capable of great things. As I mentioned at the beginning, both of my parents were veterans of World War II. My dad spent almost four years as a Prisoner of War. My mom remained in Washington, D.C., and served in the U.S. Navy. She and her parents and everybody else had U.S. Government–issued ration books. My mom took the trolley to the Naval base because everyone was permitted less than 15 liters (4 gallons) a week of gasoline. Today, I know people who go through that much gas and more everyday. Apparently, it was not uncommon to have your old-technology tires last only about two years. But when you needed new tires, you couldn't get them because rubber was rationed.

During the war years, there was great emphasis on salvage, what we would now call "recycling." People salvaged rubber, metal, and bones. Yes, bones, which could be used in making fertilizer and calcium carbonate–based adhesives for military vehicles and even airplanes. People, homemakers especially, were encouraged to clean and wash the bones thoroughly, tie them up, and leave them for the collection service. People pitched in, and the war was won. I often muse on how easy it would be for our society to have so much more by wasting even a little bit less. If it were a hockey game, it might look as if there were no goalie. We could glide the conservation puck in from our end of the consumption ice. Recycling still seems to me to hold great promise in the U.S. If people saw that climate change is a threat as dangerous to us today as the Axis was during that war, we'd be up and at 'em in no time.

For those of you who might be skeptical of my optimism, I have to point out that if you don't much care for regulation now, you might be in for a hard time. As climate change causes sea levels to rise, more and more people are going to get displaced. More and more people are going to want to come live where you are living—or worse, you will be among those forced to do the moving. Cities are going to need storm walls; farmers will need compensation to relocate their fields. If you think action on behalf of climate change is expensive, just wait until you see the price of inaction. Regulations will be required sooner or later, but if we wait until things reach crisis level they will be a lot more onerous. There may be requirements to restrict your use of gasoline. Requirements that restrict your access to proteins, such as steak and fish. Regulators watching what you put in the trash. There may be limits on shipping and air travel. And by then, your neighbors will probably be voting for these regulations. The environmental and just plain cash-money costs will be staggering the longer we go without getting going.

The more time that goes by with us screwing around not doing anything about climate change, and especially not leading the world in addressing the short- and medium-term consequences of climate change, the more The Man is going to be in your business. But it doesn't have to be this way. You could participate. You could be The Man—or The Woman. You can help lead. Together we can change the world.

# 35

# THE UNSTOPPABLE SPECIES

When I decided to write this book, I did it with one enormous goal in mind: I want to help change the world. I don't want to scare people. I don't want to blame people. I don't want to send people into despair. Those things don't enable change; they prevent it. If you're not optimistic, you will not accomplish much. At the same time, I have no illusions about what lies ahead. We have work to do—a lot of work. The world is built on a lot of old infrastructure. People and businesses have set ways of doing things, and it is not easy to get them to adapt to the new reality. But change is vital.

We have to transform the way we produce, transmit, and use energy. We have to transform the way we live, without moving backward in how well we live. With those changes will come enormous opportunities for manufacturers, inventors, venture capitalists, and entrepreneurs. Entire new industries may emerge to provide better ways of storing energy, especially electrical energy. If you and your colleagues can find a way to desalinate seawater at a much lower energy cost than we do now, you will improve the lives of billions of humans—

and uncountable numbers of other species. You will also probably become quite wealthy in the process. For most of the people on this planet, though, the payoff will be far more subtle. We will enjoy cleaner air and water. We will benefit, invisibly, from bad things that do not happen: extreme weather, sea-level rise, oceans becoming more acidic. To some extent, those things are coming no matter what we do, but we have the chance—right now—to make them much milder than they will be if we do nothing.

To confront climate change, we all have to embrace two ideas. They are simple and familiar ideas, but that does not make them any less true. First: We are all in this together. Everyone you will ever meet is from here—Earth. It is everybody's house, and it is everybody's home. Let's work as a team to make our home as clean, comfortable, and sustainable as possible. Second: The longest journey begins with a single step. Every little and big thing we do to address the warming of our world and its shifting weather patterns will help everyone on Earth. The sooner we get to work, the better.

We can take steps big and small to educate people everywhere, especially girls and women. As we do, we will have a great many more good ideas and skilled workers to address the looming climate changes. As is so popular to say these days, "Do the math." Half the humans are women and girls, so half the problem-solving engineers, researchers, technicians, and artisans could be women, too. Right now women are greatly underrepresented in most areas of science and technology. By adding girls and women to the science- and technology-educated workforce, we will nearly double the number of minds and hands available to fix things and craft new policies. We will have twice as many skilled workers and thinkers to address the looming climate changes. Let's give all the people of the Next Great Generation the chance to change the world. That's the practical aspect of diligent effort worldwide to provide educational opportunity for girls and women.

There is also another question of fairness here. It is just plain wrong to exclude girls from opportunity. And wait, there's one more enormous thing. Educated women have fewer babies. This is a universal demographic trend. The kids they do raise have more resources; they are wanted and loved. They have an opportunity to thrive. If we embrace this idea and support education for everyone worldwide, the human population will stabilize and eventually, sustainably decrease. Having fewer humans here in, say, 2115 will put a lighter load on Earth's ecosystems.

This is all doable. We just have to decide it's worth doing and support education for everyone everywhere. That long journey can start with the single step of connecting everyone on Earth to the Internet. If everyone everywhere had access to the world's information, and learned the critical thinking skills needed to access and sort that information, we would change the world quickly.

In the last chapter I mentioned the idea of introducing a carbon-production fee of ten dollars a ton. It's a way of institutionalizing the idea that we are all in this together, that we all need to be good home-owners to our planet. That level of carbon fee is only a beginning—a first step. Over the course of fifteen years or so, we might raise it to forty or fifty dollars a ton. Such a scheme would change the world in a very good way. You would see new costs in your life, but you would also see new savings. It's the same as if you insulate your home, or repair a rotting roof. There is a price to be paid for maintaining and improving your home; there is a greater price to be paid for not doing it. And only one option leads to a sustainable, unstoppable future.

On the technology side, we still have a lot of work to do to address climate change. On the policy side, we already have the essential tools. We know how to assess the cost of carbon emissions in a fair way, and we have the means to make logical use of the wealth that a carbon fee would generate. Some of the money can be directed to crucial research and development. We do not need to come up with some crazy new bureaucracies to make this happen. We have academic, scientific, and engineering assessment systems in place already. We have the National Research Council. We have universities full of brilliant students who are ready to come up with the scientific and engineering ideas and innovations to create new sustainable clean green infrastructure. They can compete for the funding to carry out the necessary research. We can use the Department of Energy to assess which schemes are mostly likely to succeed, and we can enable start-ups and established corporations to pursue new technology by building pilot plants and experimental stations.

Furthermore, some of the carbon-fee money can be redirected for education. At least half of what we learn about science we learn informally. That's a technical, educator-official term for learning that takes place outside of the classroom. Ten years old is about as old as you can be to develop a lifelong passion for science. As an informal educator, I have come to believe it's about as old as you can be to develop a lifelong passion for anything. So we could, based on these well-researched facts, invest in informal education for very young students in the U.S. and around the world. It's in everyone's best interest—everyone's. And if we worked to include every preschool and elementary schoolkid in the world, it would not only be good; it would be fair.

Fairness is a clear and present issue here. Is it fair that large countries with barely a quarter of the world's population consume most of the energy and produce most of the greenhouse gases? Is it fair that some island nations are disappearing under the rising seas, because other continental countries have made it so? If we also consider that the industrialized countries have the ability to develop the new technologies, the new large public energy projects, and the new means to distribute wealth, then perhaps there is a way forward that is as fair as is possible for everyone. And let's be clear: Nobody wants to go backward. Nobody will stand for a declining standard of living. That is part of the call to the Next Great Generation: to make a better world for everyone.

Here in the U.S., the richest nation on Earth, along with our remarkable output of climate-changing greenhouse gases, we have a vocal minority that rejects the concept of fairness I am outlining here. We have an extraordinarily well-funded cohort of climate-change deniers who are supported or funded by the fossil-fuel industry and its deceptively named political action committees or PACs. You might hear about the activities of American Energy Alliance, American Crossroads, Citizens for a Sound Environment, or the Foundation of Research on Economics and the Environment. Please research these groups to understand their true agenda. They generally oppose doing anything about climate change. Their justification for avoiding action on climate change is the belief that any action would reduce freedom, or hurt the economy, or damage our standard of living. They believe that change is a zero-sum game: If you improve the environment, then something else must suffer.

Fundamentally, the doubters do not believe in progress. They do not believe in the ability of humans to solve problems. They do not believe even in the evidence of history. In the United States, the air and water are drastically cleaner than they were a couple of generations ago. Regulations that many doubters claimed would wreck the economy instead gave us a safer and cleaner world. In many cases the doubters used the exact same arguments we hear today. Fortunately, we as a nation pushed forward. But the improvements we've made so far are not enough. They are only a start—an important but insufficient first step.

Near as I can tell, the doubters and deniers do not want to confront the immediate reality of a warming global climate, and all the disruptions that will come with it. They seem to believe that personal freedom takes priority over any action an individual might take. That's not how my parents' generation went about winning the war. It looks to me that climate deniers are, as the psychiatrists say, in denial. The problem is too big to deal with . . . or something. From my point of view, the doubters' agenda stems not so much from mean-spiritedness as from a lack of critical thinking skills—and a lack of faith in human ingenuity. There is also a natural but troublesome human tendency to close one's eyes to a situation one does not wish to acknowledge. Clenching fists and stamping feet at the scientific evidence may feel like it is making the problem go away, but the world is still warming all the same.

For humankind to get through the coming decades, we are going to have to show the U.S. voters and taxpayers that the deniers are causing trouble, leaving our world worse than they found it. They are bad homeowners. With a greater awareness of the troubles ahead and the opportunities before us, citizens like you and me can vote the elected deniers out of office. Along with that, we'll have to work hard to ignore the strident deniers in the media.

I often muse about what a marvelous thing it would be to change a few deniers' minds. If someone has diligently put in twenty or thirty years of strident denying, changing her or his mind might seem like an impossible objective. But then I consider the example of the smoker who quit. You may have been a smoker yourself at one time. My parents were; everybody was back then. Then they quit. And in your experience, who is the most strident antismoker? Who is the person who can't stand the smell, and leaves the room? It is often someone who used to smoke. Once they quit, former smokers often become the antismoking movement's foot soldiers. So I take heart. If we chip away at the Fox News hosts, if we chip away at the large field of 2016 Republican presidential candidates, if we make it clear that climate deniers are not politically credible, maybe one or two of them will turn around, like a former smoker. The scales may drop from their eyes, and they may become part of the many solutions that are required. If one of them does turn around, let's embrace that person. His or her influence could help conserve the world for humans everywhere.

Because we have put so much carbon dioxide and other greenhouse gases into the atmosphere over the last two and half centuries, there is a great fraction of climate change that's coming no matter what we do today. A certain amount of global warming and weather disruption is baked in; in a word, it is unstoppable. But you know what? So are we. Humans, and human progress, are also unstoppable.

We have made it through millennia of disease, famine, drought, and even an ice age. Until this modern time, we humans had little control over what became of our Earth, our home. We were along for the orbital-oscillating, axis-tilting, snowcap-cycling ride. But today, we have the technical means to watch over our home planet and take responsibility for our own fate. No generation before had the weather records on the scale we have today and the computing power to crunch those numbers. No generation before us could use geospatially tracked buoys and submersibles to know the ocean's currents and ecosystems as we do now. No generation before us had the assets in space to look down and see the atmosphere as it really is, thin and vulnerable. In short, no generation before the one coming of age today could so readily be unstoppable.

Right now, today, we have among us the Next Great Generation, the one that will carry this process forward to the places that it must go. The young people, the innovators, the entrepreneurs, the engineers, and the just plain hard workers are among us. Together we will accept the challenge of creating clean energy and providing clean water to everyone on Earth. The Next Great Generation can, and will, change the world—for all of us, together.

# ACKNOWLEDGMENTS

As I often remark, everyone you'll ever meet knows something you don't. There are an essentially uncountable number of people who have shaped my thinking on climate change and energy production.

My father was a confident, competent outdoorsman. He gave me respect for the forest. My mother was good at just about everything; she emphasized academic achievement and gave me confidence to make it through dark nights. My sister nudged me at homework time. My brother showed me how to throw, run, catch, and ride. I can't imagine life without his tutelage. In more recent years, I've been influenced by the work of climate scientist Michael Mann, who developed the first hockey stick graph showing the rate the world is warming; he helped me get the facts right here. Dan Miller not only designed the first Cornell Ultimate flying disk jersey, he is the guy who showed me the wisdom of sustainable industries and the big idea of a fee-and-dividend economy. Don Prothero is some kind of photographic memory tie-it-all-together thinker. My agent, Nick Pampenella, and my assistant, Christine Sposari, kept the electronic paperwork flowing. They amaze me every day with their attention to detail and good nature. The star of this part of the show though, is Corey Powell. Without his insight, attention to detail, and sense of humor, this book would never have worked—it never would have come to be. Thank you indeed.

I also owe a debt to my friends and colleagues who explore space. Exploration of the planets Venus and Mars provided modern investigators with the evidence of climate change on other worlds, which helped lead to the discovery of climate change on ours. With the trouble ahead for so many of us dwelling here on Earth, I keep in mind that space exploration brings out the best in us. It's inherently optimistic. It draws humankind together in knowing the cosmos and our place within it. I'm grateful for the support of the tens of thousands of members of The Planetary Society, and especially my colleagues on the Board of Directors. They encouraged me to take this job. It's hard to express how much I appreciate their guidance and friendship.

I guess I've been writing this book since I was in junior high school (nowadays called middle school). I used to wear a sign on my back that said Pedals Don't PΘllute, when I rode my bike (the *o* in *Pollute* was the Earth symbol). I rode that bike to the first few Earth Days on the National Mall in Washington, D.C. I've been working for a healthier, safer world ever since. Four decades later, I finished this book. Thank you for picking it up.

Bill Nye

The Tar Sands, Fort McMurray, Alberta

•

Working with Bill Nye on this book has been a double inspiration. It has taught me a great many exciting things about the world that I didn't know before. Even better, it has drawn out a great many other exciting things that I already knew without even realizing. Put more simply: It has made me feel both enlightened and activated. I hope that the readers of *Unstoppable* walk away with that same sense of empowerment.

I am keenly aware that the message of this book could not exist without its remarkable messengers—the generations of scientists who devoted their lives to investigating the laws of nature and applying them to improve the human condition. It is impossible to reflect on their efforts and not feel optimistic about our prospects as a species. I am elevated, too, by the minds in my own home. My wife, Lisa, has been an unflagging source of support, perspective, and good humor. And my daughters, Eliza and Ava, have a delightful, seemingly innate instinct for environmental responsibility and creative problem solving. In them I see, every day, the faces of the Next Greatest Generation.

Corey S. Powell

# INDEX